Econophysics and Sociophysics: Recent Progress and Future Directions

New Economic Windows

More information about this series at http://www.springer.com/series/6901

Frédéric Abergel · Hideaki Aoyama
Bikas K. Chakrabarti · Anirban Chakraborti
Nivedita Deo · Dhruv Raina
Irena Vodenska
Editors

Econophysics and Sociophysics: Recent Progress and Future Directions

Springer

Editors
Frédéric Abergel
CentraleSupélec
Châtenay-Malabry
France

Hideaki Aoyama
Department of Physics, Graduate School of Science
Kyoto University
Kyoto
Japan

Bikas K. Chakrabarti
Saha Institute of Nuclear Physics
Kolkata
India

Anirban Chakraborti
Jawaharlal Nehru University
New Delhi
India

Nivedita Deo
Department of Physics and Astrophysics
University of Delhi
New Delhi
India

Dhruv Raina
Zakir Husain Centre for Educational Studies
Jawaharlal Nehru University
New Delhi
India

Irena Vodenska
Administrative Sciences
Metropolitan College, Boston University
Boston
USA

ISSN 2039-411X ISSN 2039-4128 (electronic)
New Economic Windows
ISBN 978-3-319-47704-6 ISBN 978-3-319-47705-3 (eBook)
DOI 10.1007/978-3-319-47705-3

Library of Congress Control Number: 2016954603

Printed on acid-free paper

This Springer imprint is published by Springer Nature
The registered company is Springer International Publishing AG
The registered company address is: Gewerbestrasse 11, 6330 Cham, Switzerland

Preface

The essays appearing in this volume were presented at the international workshop entitled "Econophys-2015" held at the Jawaharlal Nehru University and University of Delhi, New Delhi, from November 27, 2015, to December 1, 2015. The workshop commemorated two decades of the formal naming of the field called "Econophysics." Prof. H.E. Stanley (Boston University, USA) first used the word in 1995 at the Statphys-Kolkata Conference, held at Kolkata, India. Econophysics-2015 was held in continuation of the "Econophys-Kolkata" series of conferences, hosted at Kolkata at regular intervals since 2005. This event was organized jointly by Jawaharlal Nehru University, University of Delhi, Saha Institute of Nuclear Physics, CentraleSupélec, Boston University, and Kyoto University.

In this rapidly growing interdisciplinary field, the tools of statistical physics that include extracting the average properties of a macroscopic system from the microscopic dynamics of the system have proven to be useful for modeling socioeconomic systems, or analyzing the time series of empirical observations generated from complex socioeconomic systems. The understanding of the global behavior of socioeconomic systems seems to need concepts from many disciplines such as physics, computer science, mathematics, statistics, financial engineering, and the social sciences. These tools, concepts, and theories have played a significant role in the study of "complex systems," which include examples from the natural and social sciences. The social environment of many complex systems shares the common characteristics of competition, among heterogeneous interacting agents, for scarce resources and their adaptation to dynamically changing environments. Interestingly, very simple models (with a very few parameters and minimal assumptions) taken from statistical physics have been easily adapted, to gain a deeper understanding of, and model complex socioeconomic problems. In this workshop, the main focus was on the modeling and analyses of such complex socioeconomic systems undertaken by the community working in the fields of econophysics and sociophysics.

The essays appearing in this volume include the contributions of distinguished experts and their coauthors from all over the world, largely based on the presentations at the meeting, and subsequently revised in light of referees' comments. For completeness, a few papers have been included that were accepted for presentation but were not presented at the meeting since the contributors could not attend due to unavoidable reasons. The contributions are organized into three parts. The first part comprises papers on "econophysics." The papers appearing in the second part include ongoing studies in "sociophysics." Finally, an "Epilogue" discusses the evolution of econophysics research.

We are grateful to all the local organizers and volunteers for their invaluable roles in organizing the meeting, and all the participants for making the conference a success. We acknowledge all the experts for their contributions to this volume, and Shariq Husain, Arun Singh Patel, and Kiran Sharma for their help in the LaTeX compilation of the articles. The editors are also grateful to Mauro Gallegati and the Editorial Board of the New Economic Windows series of the Springer-Verlag (Italy) for their continuing support in publishing the Proceedings in their esteemed series.[1] The conveners (editors) also acknowledge the financial support from the Jawaharlal Nehru University, University of Delhi, CentraleSupélec, Institut Louis Bachelier, and Indian Council of Social Science Research. Anirban Chakraborti and Dhruv Raina specially acknowledge the support from the University of Potential Excellence-II (Project ID-47) of the Jawaharlal Nehru University.

Châtenay-Malabry, France — Frédéric Abergel
Kyoto, Japan — Hideaki Aoyama
Kolkata, India — Bikas K. Chakrabarti
New Delhi, India — Anirban Chakraborti
New Delhi, India — Nivedita Deo
New Delhi, India — Dhruv Raina
Boston, USA — Irena Vodenska
August 2016

---

[1]Past volumes:

1. Econophysics and Data Driven Modelling of Market Dynamics, Eds. F. Abergel, H. Aoyama, B. K. Chakrabarti, A. Chakraborti, A. Ghosh, New Economic Windows, Springer-Verlag, Milan, 2015.
2. Econophysics of Agent-based models, Eds. F. Abergel, H. Aoyama, B. K. Chakrabarti, A. Chakraborti, A. Ghosh, New Economic Windows, Springer-Verlag, Milan, 2014.
3. Econophysics of systemic risk and network dynamics, Eds. F. Abergel, B. K. Chakrabarti, A. Chakraborti and A. Ghosh, New Economic Windows, Springer-Verlag, Milan, 2013.
4. Econophysics of Order-driven Markets, Eds. F. Abergel, B. K. Chakrabarti, A. Chakraborti, M. Mitra, New Economic Windows, Springer-Verlag, Milan, 2011.
5. Econophysics & Economics of Games, Social Choices and Quantitative Techniques, Eds. B. Basu, B. K. Chakrabarti, S. R. Chakravarty, K. Gangopadhyay, New Economic Windows, Springer-Verlag, Milan, 2010.
6. Econophysics of Markets and Business Networks, Eds. A. Chatterjee, B. K. Chakrabarti, New Economic Windows, Springer-Verlag, Milan 2007.
7. Econophysics of Stock and other Markets, Eds. A. Chatterjee, B. K. Chakrabarti, New Economic Windows, Springer-Verlag, Milan 2006.
8. Econphysics of Wealth Distributions, Eds. A. Chatterjee, S. Yarlagadda, B. K. Chakrabarti, New Economic Windows, Springer-Verlag, Milan, 2005.

Contents

Part I
Econophysics

Chapter 1
Why Have Asset Price Properties Changed so Little in 200 Years

Jean-Philippe Bouchaud and Damien Challet

Abstract We first review empirical evidence that asset prices have had episodes of large fluctuations and been inefficient for at least 200 years. We briefly review recent theoretical results as well as the neurological basis of trend following and finally argue that these asset price properties can be attributed to two fundamental mechanisms that have not changed for many centuries: an innate preference for trend following and the collective tendency to exploit as much as possible detectable price arbitrage, which leads to destabilizing feedback loops.

1.1 Introduction

According to mainstream economics, financial markets should be both efficient and stable. Efficiency means that the current asset price is an unbiased estimator of its fundamental value (aka "right", "fair" or "true") price. As a consequence, no trading strategy may yield statistically abnormal profits based on public information. Stability implies that all price jumps can only be due to external news.

Real-world price returns have surprisingly regular properties, in particular fat-tailed price returns and lasting high- and low-volatility periods. The question is therefore how to conciliate these statistical properties, both non-trivial and universally observed across markets and centuries, with the efficient market hypothesis.

J.-P. Bouchaud
Capital Fund Management, Rue de l'Université, 23, 75007 Paris, France
e-mail: jean-philippe.bouchaud@cfm.fr

J.-P. Bouchaud
Ecole Polytechnique, Palaiseau, France

D. Challet (✉)
Laboratoire de Mathématiques et Informatique Pour la Complexité et les Systèmes, CentraleSupélec, University of Paris Saclay, Paris, France
e-mail: damien.challet@centralesupelec.fr

D. Challet
Encelade Capital SA, Lausanne, Switzerland

F. Abergel et al. (eds.), *Econophysics and Sociophysics: Recent Progress and Future Directions*, New Economic Windows,
DOI 10.1007/978-3-319-47705-3_1

The alternative hypothesis is that financial markets are intrinsically and chronically unstable. Accordingly, the interactions between traders and prices inevitably lead to price biases, speculative bubbles and instabilities that originate from feedback loops. This would go a long way in explaining market crises, both fast (liquidity crises, flash crashes) and slow (bubbles and trust crises). This would also explain why crashes did not wait for the advent of modern HFT to occur: whereas the May 6 2010 flash crash is well known, the one of May 28 1962, of comparable intensity but with only human traders, is much less known.

The debate about the real nature of financial market is of fundamental importance. As recalled above, efficient markets provide prices that are unbiased, informative estimators of the value of assets. The efficient market hypothesis is not only intellectually enticing, but also very reassuring for individual investors, who can buy stock shares without risking being outsmarted by more savvy investors.

This contribution starts by reviewing 200 years of stylized facts and price predictability. Then, gathering evidence from Experimental Psychology, Neuroscience and agent-based modelling, it outlines a coherent picture of the basic and persistent mechanisms at play in financial markets, which are at the root of destabilizing feedback loops.

1.2 Market Anomalies

Among the many asset price anomalies documented in the economic literature since the 1980s (Schwert 2003), two of them stand out:

1. The Momentum Puzzle: price returns are persistent, i.e., past positive (negative) returns predict future positive (negative) returns.
2. The Excess Volatility Puzzle: asset price volatility is much larger than that of fundamental quantities.

These two effects are not compatible with the efficient market hypothesis and suggest that financial market dynamics is influenced by other factors than fundamental quantities. Other puzzles, such as the "low-volatility" and "quality" anomalies, are also very striking, but we will not discuss them here—see Ang et al. (2009), Baker et al. (2011), Ciliberti et al. (2016), Bouchaud et al. (2016) for recent reviews.

1.2.1 Trends and Bubbles

In blatant contradiction with the efficient market hypothesis, trend-following strategies have been successful on all asset classes for a very long time. Figure 1.1 shows for example a backtest of such strategy since 1800 (Lempérière et al. 2014). The regularity of its returns over 200 years implies the presence of a permanent mechanism that makes price returns persistent.

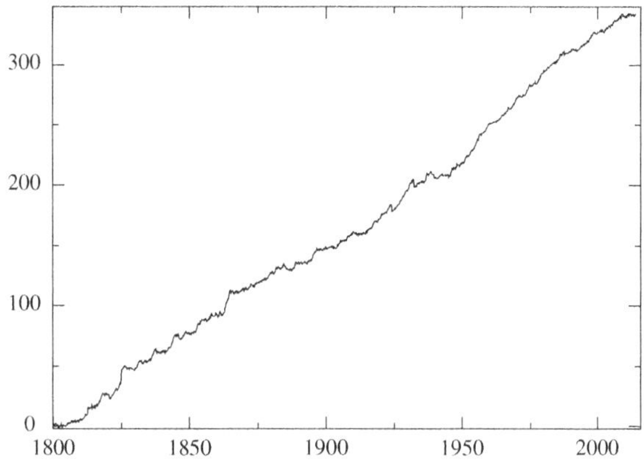

Fig. 1.1 Aggregate performance of all sectors of a trend-following strategy with the trend computed over the last six-month moving window, from year 1800 to 2013. T-statistics of excess returns is 9.8. From Lempérière et al. (2014). Note that the performance in the last 2 years since that study (2014–2015) has been strongly positive

Indeed, the propensity to follow past trends is a universal effect, which most likely originates from a behavioural bias: when faced with an uncertain outcome, one is tempted to reuse a simple strategy that seemed to be successful in the past (Gigerenzer and Goldstein 1996). The relevance of behavioural biases to financial dynamics, discussed by many authors, among whom Kahneman and Shiller, has been confirmed in many experiments on artificial markets (Smith et al. 1988), surveys (Shiller 2000; Menkhoff 2011; Greenwood and Shleifer 2013), etc. which we summarize in Sect. 1.3.

1.2.2 *Short-Term Price Dynamics: Jumps and Endogenous Dynamics*

1.2.2.1 Jump Statistics

Figure 1.2 shows the empirical price return distributions of assets from three totally different assets classes. The distributions are remarkably similar (see also Zumbach (2015)): the probability of extreme return are all $P(x) \sim |x|^{-1-\mu}$, where the exponent μ is close to 3 (Stanley et al. 2008). The same law holds for other markets (raw materials, currencies, interest rates). This implies that crises of all sizes occur and result into both positive and negative jumps, from fairly small crises to centennial crises (Figs. 1.3 and 1.4).

In addition, and quite remarkably, the probability of the occurence of price jumps is much more stable than volatility (see also Zumbach and Finger (2010)). Figure 1.4 illustrates this stability by plotting the 10-σ price jump probability as a function of time.

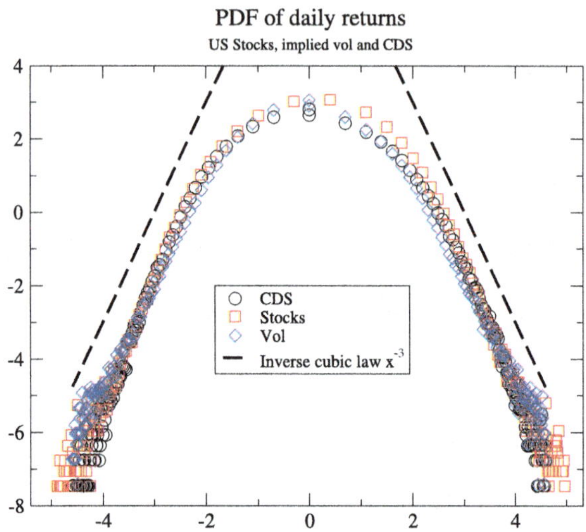

Fig. 1.2 Daily price return distributions of price, at-the-money volatility and CDS of the 283 S&P 500 that have one, between 2010 and 2013. Once one normalizes the returns of each asset class by their respective volatility, these three distributions are quite similar, despite the fact the asset classes are very different. The *dashed lines* correspond to the "inverse cubic law" $P(x) \sim |x|^{-1-3}$ (*Source* Julius Bonart)

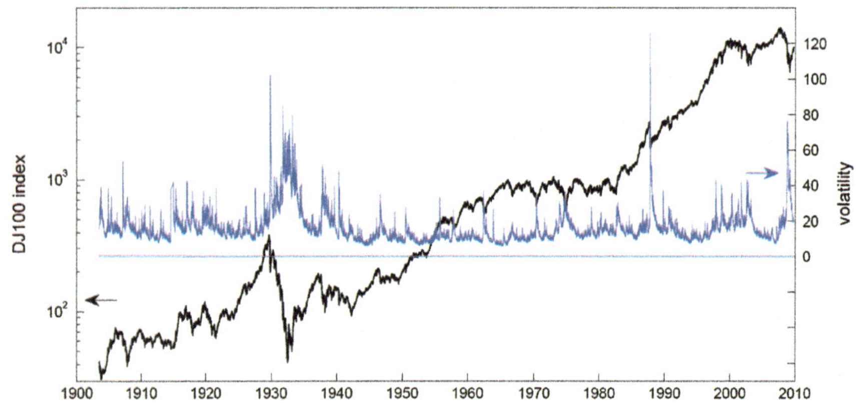

Fig. 1.3 Evolution of the Dow-Jones Industrial Average index and its volatility over a century. Sees Zumbach and Finger (2010)

1.2.2.2 The Endogenous Nature of Price Jumps

What causes these jumps? Far from being rare events, they are part of the daily routine of markets: every day, at least one 5-σ event occurs for one of the S&P500 components! According the Efficient Market Hypothesis, only some very significant pieces of information may cause large jumps, i.e., may substantially change the fundamental value of a given asset. This logical connection is disproved by empirical studies which match news sources with price returns: only a small fraction of jumps can be related to news and thus defined as an exogenous shock (Cutler et al. 1998; Fair 2002; Joulin et al. 2008; Cornell 2013).

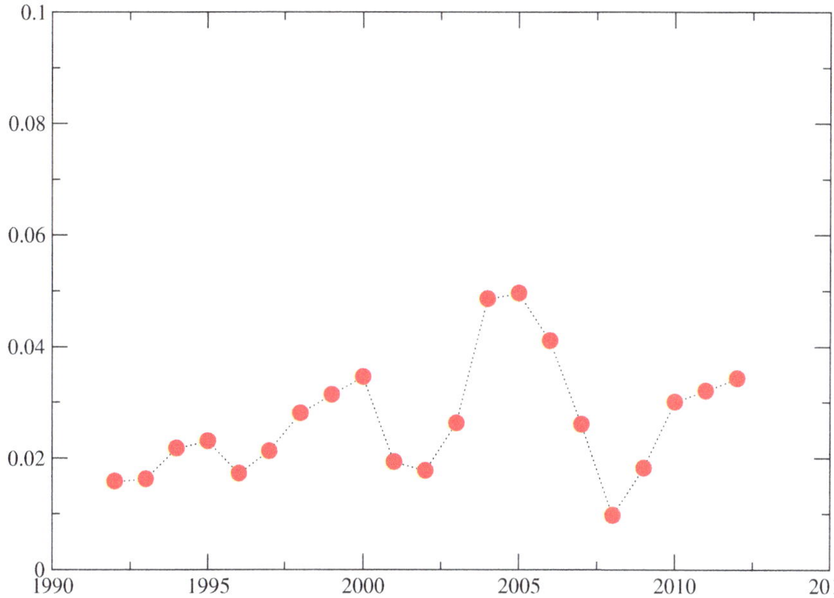

Fig. 1.4 Yearly evolution of the probability of the occurrence of 10-σ price jump for a given day for assets in the S&P500 since 1992 where σ is computed as a 250 day past average of squared daily returns. These probabilities do vary statistically from year to year, but far less than the volatility itself. This suggests that probability distributions of returns, normalized by their volatility, is universal, even in the tails (cf. also Fig. 1.3). Note that the jumps probability has not significantly increased since 1991, despite the emergence of High Frequency Trading (*Source* Stefano Ciliberti)

The inevitable conclusion is that most price jumps are self-inflicted, i.e., are endogenous. From a dynamical point of view, this means that feedback loops are so important that, at times, the state of market dynamics is near critical: small perturbations may cause very large price changes. Many different modelling frameworks yield essentially the same conclusion (Wyart et al. 2008; Marsili et al. 2009; Bacry et al. 2012; Hardiman et al. 2013; Chicheportiche and Bouchaud 2014).

The relative importance of exogenous and endogenous shocks is then linked to the propensity of the financial markets to hover near critical or unstable points. The next step is therefore to find mechanisms that systematically tend to bring financial markets on the brink.

1.3 Fundamental Market Mechanisms: Arbitrage, Behavioural Biases and Feedback Loops

In short, we argue below that greed and learning are two sufficient ingredients to explain the above stylized facts. There is no doubt that human traders have always tried to outsmart each other, and that the members the *homo sapiens sapiens* clique have some learning abilities. Computers and High Frequency Finance then merely decrease the minimum reaction speed (Hardiman et al. 2013) without modifying much the essence of the mechanisms at play.

In order to properly understand the nature of the interaction between investors in financial markets, one needs to keep two essential ingredients

1. Investor heterogeneity: the distribution of their wealth, trading frequency, computing power, etc. have heavy tails, which prevents a representative agent approach.
2. Asynchronism: the number of trades per agent in a given period is heavy-tailed, which implies that they do not trade synchronously. In addition, the continuous double auction mechanism implies sequential trading: only two orders may interact at any time.

One thus cannot assume that all the investors behave in the same way, nor that they can be split into two or three categories, which is nevertheless a common assumption when modelling or analyzing market behaviour.

1.3.1 Speculation

Although the majority of trades are of algorithmic nature nowadays, most traders (human or artificial) use the same types of strategies. Algorithmic trading very often simply implements analysis and extrapolation rules that have been used by human traders since immemorial times, as they are deeply ingrained in human brains.

1.3.1.1 Trend Following

Trend-following in essence consists in assuming that future price changes will be of the same sign as last past price changes. It is well-known that this type of strategy may destabilize prices by increasing the amplitude and duration of price excursions. Bubbles also last longer because of heavy-tailed trader heterogeneity. Neglecting new investors for the time being, the heavy-tailed nature of trader reaction times implies that some traders are much slower than others to take part to a nascent bubble. This causes a lasting positive volume imbalance that feeds a bubble for a long time. Finally, a bubble attracts new investors that may be under the impression that this bubble grow further. The neuronal processes that contribute the emergence and duration will bubbles are discussed in Sect. 1.3.4.2.

1.3.1.2 Contrarian Behaviour

Contrarian trading consists in betting on mean-reverting behavior: price excursions are deemed to be only temporary, i.e., the price will return to some reference ("fundamental" or other) value. Given the heterogeneity of traders, one may assume that they do not all have the same reference value in mind. The dynamical effects of this type of strategies is to stabilize price (with respect to its perceived reference value).

1.3.1.3 Mixing Trend Followers and Contrarians

In many simplified agent-based models (De Grauwe et al. 1993; Brock and Hommes 1998; Lux and Marchesi 1999) both types of strategies are used by some fractions of the trader populations. A given trader may either always use the same kind of strategy (Frankel et al. 1986; Frankel and Froot 1990), may switch depending on some other process (Kirman 1991) or on the recent trading performance of the strategies (Brock and Hommes (1998), Wyart and Bouchaud (2007), Lux and Marchesi (1999)). In a real market, the relative importance of a given type of strategy is not constant, which influences the price dynamics.

Which type of trading strategy dominates can be measured in principle. Let us denote the price volatility measured over a single time step by σ_1. If trend following dominates, the volatility of returns measured every T units of time, denoted by σ_T will be larger than $\sigma_1\sqrt{T}$. Conversely, if mean-reverting dominates, $\sigma_T < \sigma_1\sqrt{T}$. Variance-ratio tests, based on the quantity $\sigma_T/(\sigma_1\sqrt{T})$, are suitable tools to assess the state of the market (see Charles and Darné (2009) for a review); see for example the PUCK concept, proposed by Mizuno et al. (2007).

When trend following dominates, trends and bubbles may last for a long time. The bursting of a bubble may be seen as mean-reversion taking (belatedly) over. This view is too simplistic, however, as it implicitly assumes that all the traders have the same calibration length and the same strategy parameters. In reality, the periods of calibration used by traders to extrapolate price trends are very heterogeneous. Thus, strategy heterogeneity and the fact that traders have to close their positions some time imply that a more complex analysis is needed.

1.3.2 Empirical Studies

In order to study the behaviour of individual investors, the financial literature makes use of several types of data

1. Surveys about individual strategies and anticipation of the market return over the coming year (Shiller 2000; Greenwood and Shleifer 2013).
2. The daily investment flows in US securities of the sub-population of individual traders. The transactions of individual traders are labelled as such, without any information about the identity of the investor (Kaniel et al. 2008).
3. The daily net investment fluxes of each investor in a given market. For example, Tumminello et al. (2012) use data about Nokia in the Finish stock exchange.

4. Transactions of all individual investors of a given broker (Dorn et al. 2008; de Lachapelle and Challet 2010). The representativity of such kind of data may be however uestionned (cf. next item).
5. Transactions of all individual investors of all the brokers accessing a given market. Jackson (2004) shows that the behaviour of individual investors is the same provided that they use an on-line broker.

1.3.2.1 Trend Follower Versus Contrarian

Many surveys show that institutional and individual investors expectation about future market returns are trend-following (e.g. Greenwood and Shleifer 2013), yet the analysis of the individual investors' trading flow at a given frequency (i.e., daily, weekly, monthly) invariably point out that their actual trading is dominantly contrarian as it is anti-correlated with previous price returns, while institutional trade flow is mostly uncorrelated with recent price changes on average (Grinblatt and Keloharju (2000), Jackson (2004), Dorn et al. (2008), Lillo et al. (2008), Challet and de Lachapelle (2013)). In addition, the style of trading of a given investor only rarely changes (Lillo et al. 2008).

Both findings are not as incompatible as it seems, because the latter behaviour is consistent with price discount seeking. In this context, the contrarian nature of investment flows means that individual investors prefer to buy shares of an asset after a negative price return and to sell it after a positive price return, just to get a better price for their deal. If they neglect their own impact, i.e., if the current price is a good approximation of the realized transaction price, this makes sense. If their impact is not negligible, then the traders buy when their expected transaction price is smaller than the current price and conversely (Batista et al. 2015).

1.3.2.2 Herding Behaviour

Lakonishok et al. (1992) define a statistical test of global herding. US mutual funds do not herd, while individual investors significantly do (Dorn et al. 2008). Instead of defining global herding, Tumminello et al. (2012) define sub-groups of individual investors defined by the synchronization of their activity and inactivity, the rationale being that people that use the same way to analyse information are likely to act in the same fashion. This in fact defines herding at a much more microscopic level. The persistent presence of many sub-groups sheds a new light on herding. Using this method, Challet et al. (2016) show that synchronous sub-groups of institutional investors also exist.

1.3.2.3 Behavioural Biases

Many behavioural biases have been reported in the literature. Whereas they are only relevant to human investors, i.e., to individual investors, most institutional funds are not (yet) fully automated and resort to human decisions. We will mention two of the most relevant biases.

Human beings react different to gains and to losses (see e.g. Prospect Theory Kahneman and Tversky 1979) and prefer positively skewed returns to negatively skewed returns (aka the "lottery ticket" effect, see Lemperiere et al. 2016). This has been linked to the disposition bias, which causes investors to close too early winning trades and too late losing ones (Shefrin and Statman 1985; Odean 1998; Boolell-Gunesh et al. 2009) (see however Ranguelova 2001; Barberis and Xiong 2009; Annaert et al. 2008). An indisputable bias is overconfidence, which leads to an excess of trading activity, which diminishes the net performance (Barber and Odean 2000, see also Batista et al. 2015 for a recent experiment eliciting this effect). This explains why male traders earn less than female trades (Barber and Odean 2001). Excess confidence is also found in individual portfolios, which are not sufficiently diversified. For example, individual traders trust too much their asset selection abilities (Goetzmann and Kumar 2005; Calvet et al. 2007).

1.3.3 Learning and Market Instabilities

Financial markets force investors to be adaptive, even if they are not always aware of it (Farmer 1999; Zhang 1999; Lo 2004). Indeed, strategy selection operates in two distinct ways

1. Implicit: assume that an investor always uses the same strategy and never recalibrates its parameters. The performance of this strategy modulates the wealth of the investor, hence its relative importance on markets. In the worst case, this investor and his strategy effectively disappears. This is the argument attributed to Milton Friedman according to which only rational investors are able to survive in the long run because the uninformed investors are weeded out.
2. Explicit: investors possess several strategies and use them in an adaptive way, according to their recent success. In this case, strategies might die (i.e., not being used), but investors may survive.

The neo-classical theory assumes the convergence of financial asset prices towards an equilibrium in which prices are no longer predictable. The rationale is that market participants are learning optimally such that this outcome is inevitable. A major problem with this approach is that learning requires a strong enough signal-to-noise ratio (Sharpe ratio); as the signal fades away, so does the efficiency of any learning scheme. As a consequence, reaching a perfectly efficient market state is impossible in finite time.

This a major cause of market instability. Patzelt and Pawelzik (2011) showed that optimal signal removal in presence of noise tends to converge to a critical state characterized by explosive and intermittent fluctuations, which precisely correspond to the stylized facts described in the first part of this paper. This is a completely generic result and directly applies to financial markets. Signal-to-noise mediated transitions to explosive volatility is found in agent-based models in which predictability is measurable, as in the Minority Game (Challet and Marsili 2003; Challet et al. 2005) and more sophisticated models (Giardina and Bouchaud 2003).

1.3.4 Experiments

1.3.4.1 Artificial Assets

In their famous work, Smith et al. (1988) found that price bubbles emerged in most experimental sessions, even if only three or four agents were involved. This means that financial bubble do not need very many investors to appear. Interestingly, the more experienced the subjects, the less likely the emergence of a bubble.

More recently, Hommes et al. (2005) observed that in such experiments, the resulting price converges towards the rational price either very rapidly or very slowly or else with large oscillations. Anufriev and Hommes (2009) assume that the subjects dynamically use very simple linear price extrapolation rules (among which trend-following and mean-reverting rules),

1.3.4.2 Neurofinance

Neurofinance aims at studying the neuronal process involved in investment decisions (see Lo 2011 for an excellent review). One of the most salient result is that, expectedly, human beings spontaneously prefer to follow perceived past trends.

Various hormones play a central role in the dynamics of risk perception and reward seeking, which are major sources of positive and negative feedback loops in Finance. Even better, hormone secretion by the body modifies the strength of feedback loops dynamically, and feedback loops interact between themselves. Some hormones have a feel-good effect, while other reinforce to risk aversion.

Coates and Herbert (2008) measured the cortisol (the "stress hormone") concentration in saliva samples of real traders and found that it depends on the realized volatility of their portfolio. This means that a high volatility period durable increases the cortisol level of traders, which increases risk aversion and reduces activity and liquidity of markets, to the detriment of markets as a whole.

Reward-seeking of male traders is regulated by testosterone. The first winning round-trip leads to an increase of the level testosterone, which triggers the production of dopamine, a hormone related to reward-seeking, i.e., of another positive round-trip in this context. This motivates the trader to repeat or increase his pleasure by taking additional risk. At relatively small doses, this exposure to reward and reward-seeking has a positive effect. However, quite clearly, it corresponds to a destabilizing feedback loop and certainly reinforces speculative bubbles. Accordingly, the trading performance of investors is linked to their dopamine level, which is partly determined by genes (Lo et al. 2005; Sapra et al. 2012).

Quite remarkably, the way various brain areas are activated during the successive phases of speculative bubbles has been investigated in detail. Lohrenz et al. (2007) suggest a neurological mechanism which motivates investors to try to ride a bubble: they correlate the activity of a brain area with how much gain opportunities a trader has missed since the start of a bubble. This triggers the production of dopamine, which in turn triggers risk taking, and therefore generates trades. In other words, regrets or "fear of missing out" lead to trend following.

After a while, dopamine, i.e., gut feelings, cannot sustain bubbles anymore as its effect fades. Another cerebral region takes over; quite ironically, it is one of the more rational ones: DeMartino et al. (2013) find a correlation between the activation level of an area known to compute a representation of the mental state of other people, and the propensity to invest in a pre-existing bubble. These authors conclude that investors make up a rational explanation about the existence of the bubble ("others cannot be wrong") which justifies to further invest in the bubble. This is yet another neurological explanation of our human propensity to trend following.

1.4 Conclusion

Many theoretical arguments suggest that volatility bursts may be intimately related to the quasi-efficiency of financial markets, in the sense that predicting them is hard because the signal-to-noise ratio is very small (which does not imply that the prices are close to their "fundamental" values). Since the adaptive behaviour of investors tends to remove price predictability, which is the signal that traders try to learn, price dynamics becomes unstable as they then base their trading decision on noise only (Challet et al. 2005; Patzelt and Pawelzik 2011). This is a purely endogenous phenomenon whose origin is the implicit or explicit learning of the value of trading strategies, i.e., of the interaction between the strategies that investors use. This explains why these stylized facts have existed for at least as long as financial historical data exists. Before computers, traders used their strategies in the best way they could. Granted, they certainly could exploit less of the signal-to-noise ratio than we can today. This however does not matter at all: efficiency is only defined with respect to the set of strategies one has in one's bag. As time went on, the computational power increased tremendously, with the same result: unstable prices and bursts of volatility. This is why, unless exchange rules are dramatically changed, there is no reason to expect financial markets will behave any differently in the future.

Similarly, the way human beings learn also explains why speculative bubbles do not need rumour spreading on internet and social networks in order to exist. Looking at the chart of an asset price is enough for many investors to reach similar (and hasty) conclusions without the need for peer-to-peer communication devices (phones, emails, etc.). In short, the fear of missing out is a kind of indirect social contagion.

Human brains have most probably changed very little for the last two thousand years. This means that the neurological mechanisms responsible for the propensity to invest in bubbles are likely to influence the behaviour of human investors for as long as they will be allowed to trade.

From a scientific point of view, the persistence of all the above mechanisms justifies the quest for the fundamental mechanisms of market dynamics. We believe that the above summary provides a coherent picture of how financial markets have worked for at least two centuries (Reinhart and Rogoff 2009) and why they will probably continue to stutter in the future.

References

Andrew Ang, Robert J Hodrick, Yuhang Xing, and Xiaoyan Zhang. High idiosyncratic volatility and low returns: International and further us evidence. *Journal of Financial Economics*, 91(1):1–23, 2009.

Jan Annaert, Dries Heyman, Michele Vanmaele, and Sofieke Van Osselaer. Disposition bias and overconfidence in institutional trades. Technical report, Working Paper, 2008.

M Anufriev and C Hommes. Evolutionary selection of individual expectations and aggregate outcomes. *CeNDEF Working Paper University of Amsterdam*, 9, 2009.

Emmanuel Bacry, Khalil Dayri, and Jean-Francois Muzy. Non-parametric kernel estimation for symmetric Hawkes processes. application to high frequency financial data. *The European Physical Journal B*, 85(5):1–12, 2012.

Malcolm Baker, Brendan Bradley, and Jeffrey Wurgler. Benchmarks as limits to arbitrage: Understanding the low-volatility anomaly. *Financial Analysts Journal*, 67(1), 2011.

Brad M Barber and Terrance Odean. Trading is hazardous to your wealth: The common stock investment performance of individual investors. *The Journal of Finance*, 55(2):773–806, 2000.

Brad M Barber and Terrance Odean. Boys will be boys: Gender, overconfidence, and common stock investment. *The Quarterly Journal of Economics*, 116(1):261–292, 2001.

Nicholas Barberis and Wei Xiong. What drives the disposition effect? an analysis of a long-standing preference-based explanation. *the Journal of Finance*, 64(2):751–784, 2009.

Joao da Gama Batista, Domenico Massaro, Jean-Philippe Bouchaud, Damien Challet, and Cars Hommes. Do investors trade too much? a laboratory experiment. *arXiv preprint* arXiv:1512.03743, 2015.

S Boolell-Gunesh, Marie-Hélène Broihanne, and Maxime Merli. Disposition effect, investor sophistication and taxes: some French specificities. *Finance*, 30(1):51–78, 2009.

Jean-Philippe Bouchaud, Ciliberti Stefano, Augustin Landier, Guillaume Simon, and David Thesmar. The excess returns of 'quality' stocks: A behavioral anomaly. *J. Invest. Strategies*, Volume 5, Number 3 (June 2016) Pages: 51–61.

W.A. Brock and C.H. Hommes. Heterogeneous beliefs and routes to chaos in a simple asset pricing model. *Journal of Economic dynamics and Control*, 22(8-9):1235–1274, 1998.

Laurent E. Calvet, John Y. Campbell, and Paolo Sodini. Down or out: Assessing the welfare costs of household investment mistakes. *Journal of Political Economy*, 115(5):pp. 707–747, 2007. URL http://www.jstor.org/stable/10.1086/524204.

Damien Challet and Matteo Marsili. Criticality and finite size effects in a realistic model of stock market. *Phys. Rev. E*, 68:036132, 2003.

D. Challet, R. Chicheportiche, and M. Lallouache. Trader lead-lag networks and internal order crossing. 2016. in preparation.

Damien Challet and David Morton de Lachapelle. A robust measure of investor contrarian behaviour. In *Econophysics of Systemic Risk and Network Dynamics*, pages 105–118. Springer, 2013.

Damien Challet, Matteo Marsili, and Yi-Cheng Zhang. *Minority Games*. Oxford University Press, Oxford, 2005.

Amélie Charles and Olivier Darné. Variance-ratio tests of random walk: an overview. *Journal of Economic Surveys*, 23(3):503–527, 2009.

Rémy Chicheportiche and Jean-Philippe Bouchaud. The fine-structure of volatility feedback. *Physica A*, 410:174–195, 2014.

Stefano Ciliberti, Yves Lempérière, Alexios Beveratos, Guillaume Simon, Laurent Laloux, Marc Potters, and Jean-Philippe Bouchaud. Deconstructing the low-vol anomaly. *to appear in J. Portfolio Management*, 2016.

John M Coates and Joe Herbert. Endogenous steroids and financial risk taking on a London trading floor. *Proceedings of the national academy of sciences*, 105(16):6167–6172, 2008.

Bradford Cornell. What moves stock prices: Another look. *The Journal of Portfolio Management*, 39(3):32–38, 2013.

David M Cutler, James M Poterba, and Lawrence H Summers. What moves stock prices? *Bernstein, Peter L. and Frank L. Fabozzi*, pages 56–63, 1998.

Paul De Grauwe, Hans Dewachter, and Mark Embrechts. Exchange rate theory: chaotic models of foreign exchange markets. 1993.

David Morton de Lachapelle and Damien Challet. Turnover, account value and diversification of real traders: evidence of collective portfolio optimizing behavior. *New J. Phys*, 12:075039, 2010.

Benedetto DeMartino, John P. O'Doherty, Debajyoti Ray, Peter Bossaerts, and Colin Camerer. In the Mind of the Market: Theory of Mind Biases Value Computation during Financial Bubbles. *Neuron*, 80:1102, 2013. doi:10.1016/j.neuron.2013.11.002.

Daniel Dorn, Gur Huberman, and Paul Sengmueller. Correlated trading and returns. *The Journal of Finance*, 63(2):885–920, 2008.

Ray C Fair. Events that shook the market. *Journal of Business*, 75:713–732, 2002.

Doyne Farmer. Market force, ecology and evolution. Technical Report 98-12-117, Santa Fe Institute, 1999.

Jeffrey A Frankel and Kenneth A Froot. Chartists, fundamentalists, and trading in the foreign exchange market. *The American Economic Review*, 80(2):181–185, 1990.

Jeffrey A Frankel, Kenneth A Froot, et al. Understanding the us dollar in the eighties: the expectations of chartists and fundamentalists. *Economic record*, 62(1):24–38, 1986.

Irene Giardina and Jean-Philippe Bouchaud. Crashes and intermittency in agent based market models. *Eur. Phys. J. B*, 31:421–437, 2003.

Gerd Gigerenzer and Daniel G Goldstein. Reasoning the fast and frugal way: models of bounded rationality. *Psychological review*, 103(4):650, 1996.

William N Goetzmann and Alok Kumar. Why do individual investors hold under-diversified portfolios? Technical report, Yale School of Management, 2005.

Robin Greenwood and Andrei Shleifer. Expectations of returns and expected returns. *Rev. Fin. Studies*, 2013. to appear.

Mark Grinblatt and Matti Keloharju. The investment behavior and performance of various investor types: a study of Finland's unique data set. *Journal of Financial Economics*, 55(1):43–67, 2000.

Stephen J. Hardiman, Nicolas Bercot, and Jean-Philippe Bouchaud. Critical reflexivity in financial markets: a hawkes process analysis. *The European Physical Journal B*, 86(10):1–9, 2013. doi:10.1140/epjb/e2013-40107-3.

Cars Hommes, Joep Sonnemans, Jan Tuinstra, and Henk Van de Velden. Coordination of expectations in asset pricing experiments. *Review of Financial Studies*, 18(3):955–980, 2005.

Jackson. The aggregate behaviour of individual investors. 2004.

Armand Joulin, Augustin Lefevre, Daniel Grunberg, and Jean-Philippe Bouchaud. Stock price jumps: news and volume play a minor role. *Wilmott Mag. Sept/Oct*, 2008.

Daniel Kahneman and Amos Tversky. Prospect theory: an analysis of decision under risk. *Econometrica*, 47:263, 1979.

Ron Kaniel, Gideon Saar, and Sheridan Titman. Individual investor trading and stock returns. *The Journal of Finance*, 63(1):273–310, 2008.

Alan Kirman. Epidemics of opinion and speculative bubbles in financial markets. *Money and financial markets*, pages 354–368, 1991.

Josef Lakonishok, Andrei Shleifer, and Robert W Vishny. The impact of institutional trading on stock prices. *Journal of financial economics*, 32(1):23–43, 1992.

Yves Lempérière, Cyril Deremble, Trung-Tu Nguyen, Philip Andrew Seager, Marc Potters, and Jean-Philippe Bouchaud. Risk premia: Asymmetric tail risks and excess returns. *to appear in Quantitative Finance*, 2016.

Yves Lempérière, Philip Seager, Marc Potters, and Jean-Philippe Bouchaud. Two centuries of trend following. Technical report, 2014.

Fabrizio Lillo, Esteban Moro, Gabriella Vaglica, and Rosario Mantegna. Specialization and herding behavior of trading firms in a financial market. *New Journal of Physics*, 10:043019, 2008.

Andrew W Lo. The adaptive markets hypothesis. *The Journal of Portfolio Management*, 30(5): 15–29, 2004.

Andrew W Lo. Fear, greed, and financial crises: a cognitive neurosciences perspective. *by J. Fouque, and J. Langsam. Cambridge University Press, Cambridge, UK*, 2011.

Andrew W Lo, Dmitry V Repin, and Brett N Steenbarger. Fear and greed in financial markets: A clinical study of day-traders. *American Economic Review*, 95(2):352–359, 2005.

Terry Lohrenz, Kevin McCabe, Colin F Camerer, and P Read Montague. Neural signature of fictive learning signals in a sequential investment task. *Proceedings of the National Academy of Sciences of the United States of America*, 104(22):9493–9498, 2007.

Thomas Lux and Michele Marchesi. Scaling and criticality in a stochastic multi-agent model of a financial market. *Nature*, 397:498–500, 1999.

Matteo Marsili, Giacomo Raffaelli, and Benedicte Ponsot. Dynamic instability in generic model of multi-assets markets. *Journal of Economic Dynamics and Control*, 33(5):1170–1181, 2009.

Lukas Menkhoff. Are momentum traders different? implications for the momentum puzzle. *Applied Economics*, 43(29):4415–4430, 2011.

Takayuki Mizuno, Hideki Takayasu, and Misako Takayasu. Analysis of price diffusion in financial markets using puck model. *Physica A: Statistical Mechanics and its Applications*, 382(1):187–192, 2007.

Terrance Odean. Are investors reluctant to realize their losses? *The Journal of finance*, 53(5): 1775–1798, 1998.

Felix Patzelt and Klaus Pawelzik. Criticality of adaptive control dynamics. *Phys. Rev. Lett.*, 107(23):238103, 2011.

Elena Ranguelova. Disposition effect and firm size: New evidence on individual investor trading activity. *Available at SSRN 293618*, 2001.

Carmen M Reinhart and Kenneth Rogoff. *This time is different: Eight centuries of financial folly*. Princeton University Press, 2009.

Steve Sapra, Laura E Beavin, and Paul J Zak. A combination of dopamine genes predicts success by professional wall street traders. *PloS one*, 7(1):e30844, 2012.

G William Schwert. Anomalies and market efficiency. *Handbook of the Economics of Finance*, 1:939–974, 2003.

Hersh Shefrin and Meir Statman. The disposition to sell winners too early and ride losers too long: Theory and evidence. *The Journal of finance*, 40(3):777–790, 1985.

Robert J Shiller. Measuring bubble expectations and investor confidence. *The Journal of Psychology and Financial Markets*, 1(1):49–60, 2000.

Vernon L Smith, Gerry L Suchanek, and Arlington W Williams. Bubbles, crashes, and endogenous expectations in experimental spot asset markets. *Econometrica: Journal of the Econometric Society*, pages 1119–1151, 1988.

H Eugene Stanley, Vasiliki Plerou, and Xavier Gabaix. A statistical physics view of financial fluctuations: Evidence for scaling and universality. *Physica A: Statistical Mechanics and its Applications*, 387(15):3967–3981, 2008.

Michele Tumminello, Fabrizio Lillo, Jyrki Piilo, and R.N. Mantegna. Identification of clusters of investors from their real trading activity in a financial market. *New Journal of Physics*, 14:013041, 2012.

Matthieu Wyart and Jean-Philippe Bouchaud. Self-referential behaviour, overreaction and conventions in financial markets. *Journal of Economic Behavior & Organization*, 63(1):1–24, 2007.

Matthieu Wyart, Jean-Philippe Bouchaud, Julien Kockelkoren, Marc Potters, and Michele Vettorazzo. Relation between bid–ask spread, impact and volatility in order-driven markets. *Quantitative Finance*, 8(1):41–57, 2008.

Yi-Chen Zhang. Towards a theory of marginally efficient markets. *Physica A*, 269:30, 1999.

Gilles Zumbach. Cross-sectional universalities in financial time series. *Quantitative Finance*, 15(12):1901–1912, 2015.

Gilles Zumbach and Christopher Finger. A historical perspective on market risks using the DJIA index over one century. *Wilmott Journal*, 2(4):193–206, 2010.

Chapter 2
Option Pricing and Hedging with Liquidity Costs and Market Impact

F. Abergel and G. Loeper

Abstract We study the influence of taking liquidity costs and market impact into account when hedging a contingent claim. In the continuous time setting and under the assumption of perfect replication, we derive a fully non-linear pricing partial differential equation, and characterize its parabolic nature according to the value of a numerical parameter interpreted as a *relaxation coefficient* for market impact. We also investigate the case of stochastic volatility models with pseudo-optimal strategies.

2.1 Introduction

2.1.1 Position of the Problem

There is a long history of studying the effect of transaction costs and liquidity costs in the context of derivative pricing and hedging. Transaction costs due to the presence of a Bid-Ask spread are well understood in discrete time, see (Lamberton et al. 1997). In continuous time, they lead to quasi-variational inequalities, see e.g. (Zakamouline 2006), and to imperfect claim replication due to the infinite cost of hedging continuously over time. In this work, the emphasis is put rather on **liquidity costs**, that is, the extra price one has to pay over the theoretical price of a tradable asset, due to the finiteness of available liquidity at the best possible price. A reference work for the modelling and mathematical study of liquidity in the context of a dynamic hedging strategy is (Cetin et al. 2004), see also (Roch 2009), and our results can be seen as partially building on the same approach.

F. Abergel (✉)
Laboratory MICS, CentraleSupélec, 92290 Châtenay-Malabry, France
e-mail: frederic.abergel@centralesupelec.fr

G. Loeper
Mathematical Sciences, Monash University, Clayton, VIC 3800, Australia
e-mail: gregoire.loeper@monash.edu

F. Abergel et al. (eds.), *Econophysics and Sociophysics: Recent Progress and Future Directions*, New Economic Windows,
DOI 10.1007/978-3-319-47705-3_2

It is however unfortunate that a major drawback occurs when adding liquidity costs: as can easily be seen in (Cetin et al. 2004; Millot and Abergel 2011; Roch 2009), the pricing and hedging equation are not unconditionally parabolic anymore. Note that this sometimes dramatic situation can already be inferred from the early heuristics in Leland (1985): the formula suggested by Leland makes perfectly good sense for small perturbation of the initial volatility, but is meaningless when the modified volatility becomes negative. An answer to this problem is proposed in Çetin et al. (2010), where the authors introduce *super-replicating* strategies and show that the minimal cost of a super-replicating strategy solves a well-posed parabolic equation. In such a case, a perfectly replicating strategy, provided that it exists, may not be the *optimal* strategy, as there may exist a strategy with cheaper initial wealth that super-replicates the payoff at maturity. It appears however that such a situation, where liquidity costs lead to an imperfect replication, is dependent on the assumption one is making regarding the **market impact** of the delta-hedger, as some recent work of one of the author (Loeper 2013) already shows. In this work, we provide necessary and sufficient conditions that ensure the parabolicity of the pricing equation and hence, the existence and uniqueness of a self-financing, perfectly replicating strategy—at least in the complete market case.

Motivated by the need for quantitative approaches to algorithmic trading, the study of market impact in order-driven markets has become a very active research subject in the past decade. In a very elementary way, there always is an instantaneous market impact—termed *virtual impact* in Weber and Rosenow (2005)—whenever a transaction takes place, in the sense that the best available price immediately following a transaction may be modified if the size of the transaction is larger than the quantity available at the best limit in the order book. As many empirical works show, see e.g. (Almgren et al. 2005; Weber and Rosenow 2005), a relaxation phenomenon then takes place: after a trade, the instantaneous impact decreases to a smaller value, the permanent impact. This phenomenon is named **resilience** in Weber and Rosenow (2005), it can be interpreted as a rapid, negatively correlated response of the market to large price changes due to liquidity effects. In the context of derivative hedging, it is clear that there are realistic situations—e.g., a large option on an illiquid stock—where the market impact of an option hedging strategy is significant. This situation has already been addressed by several authors, see in particular (Schönbucher and Wilmott 2000; Frey and Stremme 1997; Frey 1998; Platen and Schweizer 1998), where various hypothesis on the dynamics, the market impact and the hedging strategy are proposed and studied. One may also refer to (Liu and Yong 2005; Roch 2009) for more recent related works. It is however noteworthy that in these references, liquidity costs and market impact are not considered jointly, whereas in fact, the latter is a rather direct consequence of the former. As we shall demonstrate, the level of permanent impact plays a fundamental role in the well-posedness of the pricing and hedging equation, a fact that was overlooked in previous works on liquidity costs and impact. Also, from a practical point of view, it seems relevant to us to relate the well-posedness of the modified Black-Scholes equation to a parameter that can be measured empirically using high frequency data.

2.1.2 Main Results

This paper aims at contributing to the field by laying the grounds for a reasonable yet complete model of liquidity costs and market impact for derivative hedging. Liquidity costs are modelled by a simple, stationary order book, characterized by its shape around the best price, and the permanent market impact is measured by a numerical parameter γ, $0 \leqslant \gamma \leqslant 1$: $\gamma = 0$ means no permanent impact, so the order book goes back to its previous state after the transaction is performed, whereas $\gamma = 1$ means no relaxation, the liquidity consumed by the transaction is shifted around the final transaction price. This simplified representation of market impact rests on the realistic hypothesis that the characteristic time of the derivative hedger, although comparable to, may be different from the relaxation time of the order book.

What we consider as our main result is Theorem 2.1, which states that, in the complete market case, the range of parameter for which the pricing equation is unconditionally parabolic is $\frac{2}{3} \leqslant \gamma \leqslant 1$. This result, which we find quite nice in that it is explicit in terms of the parameter γ, gives necessary and sufficient conditions for the perfectly replicating strategy to be optimal. It also sheds some interesting light on the ill-posedness of the pricing equations in the references (Cetin et al. 2004; Millot and Abergel 2011) corresponding to the case $\gamma = 0$, or (Liu and Yong 2005) corresponding to the case $\gamma = \frac{1}{2}$ within our formulation. In particular, Theorem 2.1 implies that when re-hedging occurs at the same frequency as that at which liquidity is provided to the order book—that is, when $\gamma = 1$—the pricing equation is well-posed. Note that there are some recent empirical evidence (Bershova and Rakhlin 2013) as well as a theoretical justification (Farmer et al. 2013) of the fact that the level of permanent impact should actually be equal to $\frac{2}{3}$, in striking compliance with the constraints Theorem 2.1 imposes!

It is of course interesting and important to thoroughly address the case where this condition is violated. If this is the case, see Sect. 2.8.1, one can build an option portfolio having the following property: there exist two european-style claims with terminal payoffs ϕ_1, ϕ_2 such that $\phi_1 \leqslant \phi_2$ but the perfect replication price of ϕ_1 is strictly greater than that of ϕ_2. The way out of this paradox should be via an approach similar to that developed in (Çetin et al. 2010), based on super-replication, but the situation is made much more complicated by the fact that, in our model, the dynamics is modified by the strategy, a feature not present in Çetin et al. (2010). We do find it interesting however that the perfect replication is not optimal, and are intrigued by a market where the value of γ would lead to imperfect replication.

Another interesting question is the comparison between our approach and that of (Almgren 2012), where the delta-hedging strategy of a large option trader is addressed. We want to point out that the two problems are tackled under very different sets of hypotheses: essentially, we consider strategies with infinite variation, whereas (Almgren 2012) refers on the contrary, to strategies with bounded variation. From a physical point of view, we deal with re-hedging that occurs at roughly the same frequency as that of the arrival of liquidity in the book, whereas (Almgren 2012) considers two different time scales, a slow one for the change in the optimal delta, and a fast one for the execution strategy. Hence, our results and models are significantly different.

The paper is organized as follows: after recalling some classical notations and concepts, Sect. 2.4 presents the continuous time model under scrutiny. The pricing and hedging equations are then worked out and characterized in the case of a complete market, in the single asset case in Sect. 2.5, and in the multi-asset case in Sect. 2.6. Section 2.7 touches upon the case of stochastic volatility models, for which partial results are presented. Finally, a short discussion of the two main conditions for Theorem 2.1, viz market impact level and Gamma-constraint, is presented in the concluding Sect. 2.8.

2.2 Basic Notations and Definitions

To ease notations, we will assume throughout the paper that the risk-free interest rate is always 0, and that the assets pay no dividend.

2.2.1 Discrete Time Setting

The tradable asset price is modelled by a positive stochastic process $S = (S_k)_{k=0,\dots,T}$ on a probability space $(\Omega, \mathscr{F}, P)$. The process S is adapted to the filtration $(\mathscr{F}_k)_{k=0,\dots,T}$, where $\mathscr{F}_k$ denotes the $\sigma-$field of events observable up to and including time k. Moreover, $\mathscr{F}_0$ is trivial and $\mathscr{F}_T = \mathscr{F}$.

A contingent claim is a random variable H of the following form $H = \delta^H S_T + \beta^H$ with δ^H and β^H, $\mathscr{F}_T$-measurable random variables.

A trading strategy Φ is given by two stochastic processes δ and β. δ_k (resp. β_k) is the amount of stock (resp. cash) held during period k, $(= [t_k, t_{k+1}))$ and is fixed at the beginning of that period, i.e. we assume that δ_k (resp. β_k) is $\mathscr{F}_k-$measurable $(k = 0, \dots, T)$.

The theoretical value of the portfolio at time k is given by

$$V_k = \delta_k S_k + \beta_k, \, (k = 1, \dots, T).$$

In order to avoid dealing with several rather involved cases, we assume that no transaction on the stock takes place at maturity: the claim will be settled with whatever position there is in stock, plus a cash adjustment to match its theoretical value (see the discussion in Lamberton et al. 1997, Sect. 4).

For the model to be specified, one must specify some integrability conditions on the various random variables just introduced, see e.g. (Millot 2012; Abergel and Millot 2011). However, since market impact is considered, the dynamics of S is not independent from that of the strategy (δ, β), so that this set of assumptions can only be verified a posteriori, once a strategy is chosen. Since our purpose is to use the discrete case as an illustrative example laying the ground for the continuous-time setting, we will not make such conditions more explicit.

2.2.2 Continuous Time Setting

In the continuous case, $(\Omega, \mathscr{F}, P)$ is a probability space with a filtration $(\mathscr{F}_t)_{0\leq t\leq T}$ satisfying the usual conditions of right-continuity and completeness. $T \in \mathbb{R}^{*+}$ denotes a fixed and finite time horizon. As before, $\mathscr{F}_0$ is trivial and $\mathscr{F}_T = \mathscr{F}$.

The risky asset $S = (S_t)_{0\leq t\leq T}$ is a strictly positive, continuous $\mathscr{F}_t$-semimartingale, and a trading strategy Φ is a pair of càdlàg and adapted processes $\delta = (\delta_t)_{0\leq t\leq T}$, $\beta = (\beta_t)_{0\leq t\leq T}$, while a contingent claim is described by a random variable H of the form $H = \delta^H S_T + \beta^H$, δ^H and β^H being $\mathscr{F}_T$−measurable random variables.

As in the discrete case, some further admissibility conditions must be imposed. One of the important consequences of our main result, Theorem 2.1, will be precisely to give sufficient conditions ensuring that *perfectly replicating* trading strategies are admissible.

2.2.3 Order Book, Liquidity Cost and Impact

Let us first emphasize that we are not pretending to use a realistic order book model here, but rather, a stylized version which can be considered a much simplified yet useful approximation of the way liquidity is provided to the market.

A stationary, symmetric order-book profile is considered around the **logarithm** of the price $\hat{S}_t$ of the asset S at a given time t **before** the option position is delta-hedged—think of $\hat{S}_t$ as a theoretical price in the absence of the option hedger. The relative density $\mu(x) \geqslant 0$ of the order book is the derivative of the function $M(x) \equiv \int_0^x \mu(t)dt \equiv$ number of shares one can buy (resp. sell) between the prices $\hat{S}_t$ and $\hat{S}_t e^x$ for positive (resp. negative) x.

This choice of representation in logarithmic scale is intended to avoid inconsistencies for large sell transactions.

The instantaneous—*virtual* in the terminology of (Weber and Rosenow 2005)—market impact of a transaction of size ε is then

$$I_{virtual}(\varepsilon) = \hat{S}_t(e^{M^{-1}(\varepsilon)} - 1), \tag{2.1}$$

it is precisely the difference between the price before and immediately after the transaction is completed.

The level of permanent impact is then measured via a parameter γ:

$$I_{permanent}(\varepsilon) = \hat{S}_t(e^{\gamma M^{-1}(\varepsilon)} - 1). \tag{2.2}$$

The actual cost of the same transaction is

$$C(\varepsilon) = \hat{S}_t \int_0^{\varepsilon} e^{M^{-1}(y)} dy. \tag{2.3}$$

Denote by κ the function M^{-1}. Since some of our results in **discrete time** depend on the simplifying assumption that κ is a linear function:

$$\kappa(\varepsilon) \equiv \lambda\varepsilon \tag{2.4}$$

for some $\lambda \in \mathbf{R}$, the computations are worked out explicitly in this setting.

$$I_{virtual}(\varepsilon) = \hat{S}_t(e^{\lambda\varepsilon} - 1), \tag{2.5}$$

$$I_{permanent}(\varepsilon) = \hat{S}_t(e^{\gamma\lambda\varepsilon} - 1), \tag{2.6}$$

and

$$C(\varepsilon) = \hat{S}_t \int_0^{\varepsilon} e^{M^{-1}(y)} dy \equiv \hat{S}_t \frac{(e^{\lambda\varepsilon} - 1)}{\lambda}. \tag{2.7}$$

This simplifying assumption is necessary for the derivation of the dynamic programming principle satisfied by local-risk minimizing strategies, see Sect. 2.3. Note however that this assumption plays no role in the continuous-time case, where the infinitesimal market impact becomes linear, see Eq. (2.24), and only the shape of the order book around 0 is relevant.

2.3 Cost Process with Market Impact in Discrete Time

In this section, we focus on the discrete time case. As said above, the order book is now assumed to be *flat*, so that κ is a linear function as in (2.4).

2.3.1 The Observed Price Dynamics

The model for the dynamics of the observed price—that is, the price S_k that the market can see at every time t_k after the re-hedging is complete—is now presented.

A natural modelling assumption is that the price moves according to the following sequence of events:

- First, it changes under the action of the "market" according to some (positive) stochastic dynamics for the theoretical price increment $\Delta\hat{S}_k$

$$\hat{S}_k \equiv S_{k-1} + \Delta\hat{S}_k \equiv S_{k-1}e^{\Delta M_k + \Delta A_k}, \tag{2.8}$$

where ΔM_k (resp. ΔA_k) is the increment of an $\mathscr{F}$-martingale (resp. an $\mathscr{F}$-predictable process).
- Then, the hedger applies some extra pressure by re-hedging her position, being thereby subject to liquidity costs and market impact as introduced in Sect. 2.2. As a consequence, the dynamics of the observed price is

$$S_k = S_{k-1}e^{\Delta M_k + \Delta A_k}e^{\gamma\lambda(\delta_k - \delta_{k-1})}. \tag{2.9}$$

Since this model is "exponential-linear"—a consequence of the assumption that κ is linear—this expression can be simplified to give

$$S_k = S_0 e^{M_k + A_k} e^{\gamma\lambda\delta_k}. \tag{2.10}$$

with the convention that M, A, δ are equal to 0 for $k = 0$.

2.3.2 Incremental Cost and Optimal Hedging Strategy

Following the approach developed in Millot and Abergel (2011), the incremental cost ΔC_k of re-hedging at time t_k is now studied. The strategy associated to the pair of processes β, δ consists in buying $\delta_k - \delta_{k-1}$ shares of the asset and rebalancing the cash account from β_{k-1} to β_k at the beginning of each hedging period $[t_k, t_{k+1})$. With the notations just introduced in Sect. 2.3.1, there holds

$$\Delta C_k = \hat{S}_k \frac{(e^{\lambda(\delta_k - \delta_{k-1})} - 1)}{\lambda} + (\beta_k - \beta_{k-1}). \tag{2.11}$$

Upon using a quadratic criterion, and under some assumptions ensuring the convexity of the quadratic risk, see e.g. (Millot and Abergel 2011), one easily derives the two (pseudo-)optimality conditions for local risk minimization

$$E(\Delta C_k | \mathscr{F}_{k-1}) = 0 \tag{2.12}$$

and

$$E((\Delta C_k)(\hat{S}_k(\gamma + (1-\gamma)e^{\lambda(\delta_k - \delta_{k-1})}))|\mathscr{F}_{k-1}) = 0,$$

where one must be careful to differentiate $\hat{S^k}$ with respect to δ_{k-1}, see (2.10).

This expression is now transformed—using the martingale condition (2.12) and the observed price (2.10)—into

$$E((\Delta C_k)(S_k e^{-\lambda\gamma(\delta_k-\delta_{k-1})}(\gamma + (1-\gamma)e^{\lambda(\delta_k-\delta_{k-1})}))|\mathscr{F}_{k-1}) = 0 \tag{2.13}$$

Equation (2.13) can be better understood—especially when passing to the continuous time limit—by introducing a modified price process accounting for the cumulated effect of liquidity costs and market impact, as in (Millot and Abergel 2011; Cetin et al. 2004). To this end, we introduce the

Definition 2.1 The supply price $\bar{S}$ is the process defined by

$$\bar{S}_0 = S_0 \tag{2.14}$$

and, for $k \geqslant 1$,

$$\bar{S}_k - \bar{S}_{k-1} = S_k e^{-\lambda\gamma(\delta_k-\delta_{k-1})}(\gamma + (1-\gamma)e^{\lambda(\delta_k-\delta_{k-1})}) - S_{k-1}. \tag{2.15}$$

Then, the orthogonality condition (2.13) is equivalent to

$$E((\Delta C_k)(\bar{S}_k - \bar{S}_{k-1})|\mathscr{F}_{k-1}) = 0. \tag{2.16}$$

It is classical—and somewhat more natural—to use the portfolio value process

$$V_k = \beta_k + \delta_k S_k, \tag{2.17}$$

so that one can then rewrite the incremental cost in (2.11) as

$$\Delta C_k = (V_k - V_{k-1}) - (\delta_k S_k - \delta_{k-1} S_{k-1}) + \hat{S}_k \frac{(e^{\lambda(\delta_k-\delta_{k-1})} - 1)}{\lambda}, \tag{2.18}$$

or equivalently

$$\Delta C_k = (V_k - V_{k-1}) - \delta_{k-1}(S_k - S_{k-1}) + S_k \left(\frac{e^{\lambda(\delta_k-\delta_{k-1})} - 1}{\lambda e^{\gamma\lambda(\delta_k-\delta_{k-1})}} - (\delta_k - \delta_{k-1}) \right). \tag{2.19}$$

To ease the notations, let us define, for $x \in \mathbf{R}$,

$$\mathbf{g}(x) \equiv \frac{e^{\lambda x} - 1}{\lambda e^{\gamma\lambda x}} - x. \tag{2.20}$$

The function $\mathbf{g}$ is smooth and satisfies

$$\mathbf{g}(0) = \mathbf{g}'(0) = 0, \ \mathbf{g}''(0) = (1-2\gamma)\lambda. \tag{2.21}$$

As a consequence, the incremental cost of implementing a hedging strategy at time t_k has the following expression

$$\Delta C_k = (V_k - V_{k-1}) - \delta_{k-1}(S_k - S_{k-1}) + S_k \mathbf{g}(\delta_k - \delta_{k-1}), \tag{2.22}$$

and Eq. (2.13) can be rewritten using the value process V and the supply price process $\bar{S}$ as

$$E((V_k - V_{k-1} - \delta_{k-1}(S_k - S_{k-1}) + S_k \mathbf{g}(\delta_k - \delta_{k-1}))(\bar{S}_k - \bar{S}_{k-1})|\mathscr{F}_{k-1}) = 0. \tag{2.23}$$

One can easily notice that Eqs. (2.12) and (2.13) reduce exactly to Eq. (2.1) in (Millot and Abergel 2011) when market impact is neglected ($\gamma = 0$) and the risk function is quadratic.

2.4 The Continuous-Time Setting

This section is devoted to the characterization of the limiting equation for the value and the hedge parameter when the time step goes to zero. Since the proofs are identical to those given in (Abergel and Millot 2011; Millot and Abergel 2011), we shall only provide formal derivations, limiting ourselves to the case of (continuous) Itō semi-martingales for the driving stochastic equations. However, in the practical situations considered in this paper, in particular those covered in Theorem 2.1, necessary and sufficient conditions are given that ensure the well-posedness in the classical sense of the strategy-dependent stochastic differential equations determining the price, value and cost processes, so that the limiting arguments can be made perfectly rigourous.

2.4.1 The Observed Price Dynamics

A first result concerns the dynamics of the observed price. Assuming that the underlying processes are continuous and taking limits in *ucp* topology, one shows that the continuous-time equivalent of (2.10) is

$$dS_t = S_t(dX_t + dA_t + \gamma\lambda d\delta_t) \tag{2.24}$$

where X is a continuous martingale and A is a continuous, predictable process of bounded variation.

Equation (2.24) is fundamental in that it contains the information on the strategy-dependent volatility of the observed price that will lead to fully non-linear parabolic pricing equation. In fact, the following result holds true:

Lemma 2.1 *Consider a hedging strategy δ which is a function of time and the observed price S at time t: $\delta_t \equiv \delta(S_t, t)$. Then, the observed price dynamics* (2.24) *can be rewritten as*

$$(1 - \gamma\lambda S_t \frac{\partial\delta}{\partial S})\frac{dS_t}{S_t} = dX_t + dA'_t, \tag{2.25}$$

where A' is another predictable, continuous process of bounded variation.

Proof Use Itō's lemma in Eq. (2.24).

2.4.2 Cost of a Strategy and Optimality Conditions

At this stage, we are not concerned with the actual optimality—with respect to local-risk minimization—of pseudo-optimal solutions, but rather, with pseudo-optimality in continuous time. Hence, we shall use Eqs. (2.12) and (2.23) as a starting point when passing to the continuous time limit.

Thanks to $\mathbf{g}'(0) = 0$, there holds the

Proposition 2.1 *The cost process of an admissible hedging strategy* (δ, V) *is given by*

$$C_t \equiv \int_0^t (dV_u - \delta dS_u + \frac{1}{2} S_u \mathbf{g}''(0) d < \delta, \delta >_u). \tag{2.26}$$

*Moreover, an admissible strategy is (pseudo-)***optimal** iff *it satisfies the two conditions*

- *C is a martingale*
- *C is orthogonal to the supply price process $\bar{S}$, with*

$$d\bar{S}_t = dS_t + S_t((1 - 2\gamma)\lambda d\delta_t + \mu d < \delta, \delta >_t) \tag{2.27}$$

and $\mu = \frac{1}{2}(\lambda^2(\gamma^3 + (1 - \gamma)^3))$.

In particular, if C is pseudo-optimal, there holds that

$$d < C, \bar{S} >_t \equiv d < V, S >_t - \delta d < S, S >_t + (1 - 2\gamma)\lambda S_t d < V, \delta >_t - \delta S_t (1 - 2\gamma)\lambda d < \delta, S >_t = 0. \tag{2.28}$$

2.5 Complete Market: The Single Asset Case

It is of course interesting and useful to fully characterize the hedging and pricing strategy in the case of a complete market. Hence, we assume in this section that the driving factor X is a one-dimensional Wiener process W and that $\mathscr{F}$ is its natural filtration, so that the increment of the observed price is simply

$$dS_t = S_t(\sigma dW_t + \gamma\lambda d\delta_t + dA_t) \tag{2.29}$$

where the "unperturbed" volatility σ is supposed to be constant. We also make the Markovian assumption that the strategy is a function of the state variable S and of time.

Under this set of assumptions, perfect replication is considered: the cost process C has to be identically 0, and Eq. (2.26) yields the two conditions

$$\frac{\partial V}{\partial S} = \delta, \tag{2.30}$$

and

$$\frac{\partial V}{\partial t} + \frac{1}{2}\left(\frac{\partial^2 V}{\partial S^2} + S_t \mathbf{g}''(0)\left(\frac{\partial^2 V}{\partial S^2}\right)^2\right)\frac{d < S, S >_t}{dt} = 0. \tag{2.31}$$

Applying Lemma 2.1 yields

$$(1 - \gamma\lambda S_t \frac{\partial \delta}{\partial S})\frac{dS_t}{S_t} = \sigma dW_t + dA'_t \tag{2.32}$$

leading to

$$\frac{d < S, S >_t}{dt} = \frac{\sigma^2 S_t^2}{(1 - \gamma\lambda S_t \frac{\partial \delta}{\partial S})^2}. \tag{2.33}$$

Hence, taking (2.30) into account, there holds

$$\frac{\partial V}{\partial t} + \frac{1}{2}\left(\frac{\partial^2 V}{\partial S^2} + \mathbf{g}''(0) S_t \left(\frac{\partial^2 V}{\partial S^2}\right)^2\right)\frac{\sigma^2 S_t^2}{(1 - \gamma\lambda S_t \frac{\partial \delta}{\partial S})^2} = 0 \tag{2.34}$$

or, using (2.30) and the identity $\mathbf{g}''(0) = (1 - 2\gamma)\lambda$:

$$\frac{\partial V}{\partial t} + \frac{1}{2}\left(\frac{\partial^2 V}{\partial S^2}\left(1 + (1 - 2\gamma)\lambda S_t \frac{\partial^2 V}{\partial S^2}\right)\right)\frac{\sigma^2 S_t^2}{(1 - \gamma\lambda S_t \frac{\partial^2 V}{\partial S^2})^2} = 0. \tag{2.35}$$

Equation (2.35) can be seen as the pricing equation in our model: any contingent claim can be perfectly replicated at zero cost, as long as one can exhibit a solution to (2.35). Consequently, of the utmost importance is the parabolicity of the pricing equation (2.35).

For instance, the case $\gamma = 1$ corresponding to a full market impact (no relaxation) yields the following equation

$$\frac{\partial V}{\partial t} + \frac{1}{2}\frac{\partial^2 V}{\partial S^2}\frac{\sigma^2 S^2}{(1 - \gamma\lambda S \frac{\partial^2 V}{\partial S^2})} = 0, \tag{2.36}$$

which can be shown to be parabolic, see (Loeper 2013). In fact, there holds the sharp result

Theorem 2.1 *Let us assume that* $\frac{2}{3} \leqslant \gamma \leqslant 1$. *Then, there holds:*

- *The non-linear backward partial differential operator*

$$V \to \frac{\partial V}{\partial t} + \frac{1}{2}\left(1 + (1 - 2\gamma)\lambda S \frac{\partial^2 V}{\partial S^2}\right)\frac{\sigma^2 S^2}{(1 - \gamma\lambda S \frac{\partial^2 V}{\partial S^2})^2}\frac{\partial^2 V}{\partial S^2} \tag{2.37}$$

  is parabolic.

- *Every European-style contingent claim with payoff Φ satisfying the terminal constraint*

$$\sup_{S\in\mathbf{R}^+}\left(S\frac{\partial^2\Phi}{\partial S^2}\right) < \frac{1}{\gamma\lambda} \tag{2.38}$$

can be perfectly replicated via a δ-hedging strategy given by the unique, smooth away from T, solution to Eq. (2.35).

Proof The parabolic nature of the operator is determined by the monotonicity with respect to p of the function

$$p \to F(p) = \frac{p(1+(1-2\gamma)p)}{(1-\gamma p)^2}. \tag{2.39}$$

A direct computation shows that $F'(p)$ has the sign of $1+(2-3\gamma)p$, so that F is globally monotonic increasing on its domain of definition whenever $\frac{2}{3} \leqslant \gamma \leqslant 1$. Now, given that the payoff satisfies the terminal constraint, some deep results on the maximum principle for the **second derivative** of the solution of nonlinear parabolic equations, see e.g. (Wang 1992a, b), ensure that the same constraint is satisfied globally for $t \leqslant T$, and therefore, (2.36) is globally well-posed. As a consequence, the stochastic differential equation determining the price of the asset has a classical, strong solution up to time T.

In order to keep this paper self-contained, we provide a straightforward proof of the maximum principle for the second derivative of V in the more general case where the volatility can be state- and time-dependent, as follows: differentiating twice (2.36) with respect to S yields the following equation

$$\frac{\partial U}{\partial t} + \frac{\partial^2}{\partial S^2}\left(\frac{\sigma^2 S}{2\lambda}F(U)\right) = 0, \tag{2.40}$$

where $U \equiv \lambda S\frac{\partial^2 V}{\partial S^2}$. Assuming for the moment that this is legitimate, we introduce a new unknown function $Z = \frac{\sigma^2 S}{2\lambda}F(U)$, so that Z is formally the solution to

$$\frac{\partial}{\partial t}\left(F^{-1}\left(\frac{2\lambda Z}{\sigma^2 S}\right)\right) + \frac{\partial^2 Z}{\partial S^2} = 0, \tag{2.41}$$

rewritten under the form

$$\frac{\partial Z}{\partial t} + \frac{\sigma^2 S}{2\lambda}F'(F^{-1}(Z))\frac{\partial^2 Z}{\partial S^2} - \frac{\frac{\partial \sigma^2}{\partial t}}{\sigma^2}Z = 0. \tag{2.42}$$

As a final change of unknown function, let us introduce $Y \equiv \frac{Z}{S}$, a solution to

$$\frac{\partial Y}{\partial t} + \frac{\sigma^2 S}{2\lambda}F'(F^{-1}(SY))\frac{\partial^2 Y}{\partial S^2} + \frac{\sigma^2 S}{\lambda}F'(F^{-1}(SY))\frac{\partial Y}{\partial S} - \frac{\frac{\partial \sigma^2}{\partial t}}{\sigma^2}Y = 0. \tag{2.43}$$

At this stage, and under the only natural and trivial assumption that the coefficient $\frac{\frac{\partial \sigma^2}{\partial t}}{\sigma^2}$ of the 0th term is bounded, one can apply the classical maximum principle for a smooth solution of (2.43): upon multiplying the unknown function Y by some exponential time-dependent function $e^{\alpha(T-t)}$, α large enough, one easily shows that a solution of (2.43) cannot have a local positive maximum or negative minimum; hence, it is uniformly bounded over any time interval $[0, T]$ if its terminal condition is. Once this a priori estimate is proven, the method of continuity allows one to obtain a unique, smooth classical solution (Ladyzhenskaya et al. 1968; Gilbarg and Trudinger 1998). Then, applying in reverse order the various changes of unknown function, one constructs the unique smooth, classical solution to the original equation (2.35), satisfying by construction the constraint (2.38) everywhere.

As a consequence, there exists a classical, strong solution to the SDE (2.38)—since the denominator is bounded away from 0—and the cost process introduced in Proposition 2.1 is well-defined, and identically 0. Hence, the perfect replication is possible.

Clearly, the constraint on the second derivative is binding, in that it is necessary to ensure the existence of the asset price itself. See however Sect. 2.8 for a discussion of other situations.

2.6 Complete Market: The Multi-asset Case

Consider a complete market described by d state variables $X = X_1, \ldots, X_d$: one can think for instance of a stochastic volatility model with $X_1 = S$ and $X_2 = \sigma$ when option-based hedging is available. Using tradable market instruments, one is able to generate d hedge ratio $\delta = \delta_1, \ldots, \delta_d$ with respect to the independent variables $X_1, \ldots, X_d$, that is, one can buy a combination of instruments whose price $P(t, X)$ satisfies

$$\partial_{X_i} P = \delta_i. \tag{2.44}$$

We now introduce two matrices, Λ_1 and Λ_2. Λ_1 accounts for the liquidity costs, so that its entry Λ^1_{ij} measures the virtual impact on Asset i of a transaction on Asset j: according to the simplified view of the order book model presented in Sect. 2.2.3, it would be natural to assume that Λ_1 is diagonal, but it is not necessary, and we will not make this assumption in the derivations that follow.

As for Λ_2, it measures the permanent impact, and need not be diagonal.

When $d = 1$, Λ_1 and Λ_2 are linked to the notations in Sect. 2.4 by

$$\Lambda_1 = \lambda S, \Lambda_2 = \gamma \lambda S.$$

Note that here, we proceed directly in the continuous time case, so that the actual shape of the order book plays a role only through its Taylor expansion around 0; hence, the use of the "linearized" impact via the matrices Λ_i.

The pricing equation is derived along the same lines as in Sect. 2.4: the dynamics of the observed price change can be written as

$$dX_t = d\hat{X}_t + dA_t + \Lambda_2 d\delta_t, \tag{2.45}$$

the d-dimensional version of (2.24).

Again, a straightforward application of Itō's formula in a Markovian setting yields the dynamics of the observed price

$$dX_t = (I - \Lambda_2 D\delta)^{-1} d\hat{X}_t + dA'_t. \tag{2.46}$$

where $D\delta$ contains the first-order terms in the differential of δ, in matrix form $(D\delta)_{ij} = \frac{\partial \delta_i}{\partial S_j}$.

Denote by V the value of the hedging portfolio. The d-dimensional version of Proposition 2.1 for the incremental cost of hedging is

$$dC_t = dV_t - \sum_{i=1}^{d} \delta_i dX_t^i + \frac{1}{2}\text{Trace}((\Lambda_1 - 2\Lambda_2)d < \delta, \delta >_t). \tag{2.47}$$

The market being complete, the perfect hedge condition $dC_t = 0$ yields the usual delta-hedging strategy

$$\frac{\partial V}{\partial X_i} = \delta_i, \tag{2.48}$$

so that one can now write $D\delta = \Gamma$, where Γ is the Hessian of V, and therefore, the pricing equation is

$$\partial_t V + \frac{1}{2}\text{Trace}\left(\Gamma \frac{d < X, X >_t}{dt}\right) = \text{Trace}\left(\Gamma\left(\Lambda_2 - \frac{1}{2}\Lambda_1\right)\Gamma \frac{d < X, X >_t}{dt}\right). \tag{2.49}$$

Using (2.46), one obtains

$$\partial_t V + \frac{1}{2}\text{Trace}\left[(\Gamma(I - (2\Lambda_2 - \Lambda_1)\Gamma))(M\Sigma M^{\mathrm{T}})\right] = 0. \tag{2.50}$$

where we have set $\Sigma = \frac{d<\hat{X},\hat{X}>_t}{dt}$, $M = (I - \Lambda_2\Gamma)^{-1}$ and M^{T} is the transpose of the matrix M.

In the particular case where $\Lambda_1 = \Lambda_2$ (i.e. no relaxation), the pricing equation becomes

$$\partial_t V + \frac{1}{2}\text{Trace}(\Gamma\Sigma((I - \Lambda\Gamma)^{-1})^{\mathrm{T}}) = 0 \tag{2.51}$$

or, after a few trivial manipulations using the symmetry of the matrices M and Γ,

$$\partial_t V + \frac{1}{2}\mathrm{Trace}(\Gamma(I - \Lambda\Gamma)^{-1}\Sigma) = 0. \tag{2.52}$$

In particular, the 1-dimensional case yields the equation already derived in Loeper (2013)

$$\partial_t V + \frac{1}{2}\frac{\Gamma}{1 - \lambda S\Gamma}S^2\sigma^2 = 0, \tag{2.53}$$

a particular case of Eq. (2.35) with $\gamma = 1$. The assessment of well-posedness in a general setting is related to the monotonicity of the linearized operator, and it may be cumbersome—if not theoretically challenging—to seek explicit conditions. In the case of full market impact $\Lambda_1 = \Lambda_2 \equiv \Lambda$, there holds the

Proposition 2.2 *Assume that the matrix Λ is symmetric. Then, Eq. (2.51) is parabolic on the connected component of* $\{\det(I - \Lambda\Gamma) > 0\}$ *that contains* $\{\Gamma = 0\}$.

Proof Let

$$F(\Gamma) = \mathrm{Trace}(\Gamma(I - \Lambda\Gamma)^{-1}\Sigma_t),$$

and

$$H(\Gamma) = \Gamma(I - \Lambda\Gamma)^{-1}.$$

Denoting by $\mathbb{S}_d^+$ the set of d-dimensional symmetric positive matrices, we need to show that for all $d\Gamma \in \mathbb{S}_d^+$, for all covariance matrix $\Sigma \in \mathbb{S}_d^+$, there holds

$$F(\Gamma + d\Gamma) \geq F(\Gamma).$$

Performing a first order expansion yields

$$H(\Gamma + d\Gamma) - H(\Gamma) = \Gamma(I - \Lambda\Gamma)^{-1}\Lambda d\Gamma(I - \Lambda\Gamma)^{-1} + d\Gamma(I - \Lambda\Gamma)^{-1} \tag{2.54}$$

$$= (\Gamma(I - \Lambda\Gamma)^{-1}\Lambda + I)d\Gamma(I - \Lambda\Gamma)^{-1}. \tag{2.55}$$

Using the elementary Lemma 2.2—stated below without proof—there immediately follows that

$$F(\Gamma + d\Gamma) - F(\Gamma) = \mathrm{Trace}((I - \Gamma\Lambda)^{-1}d\Gamma(I - \Lambda\Gamma)^{-1}\Sigma) \tag{2.56}$$

$$= \mathrm{Trace}(d\Gamma(I - \Lambda\Gamma)^{-1}\Sigma(I - \Gamma\Lambda)^{-1}). \tag{2.57}$$

Then, the symmetry condition on Λ allows to conclude the proof of Proposition 2.2.

Lemma 2.2 *The following identity holds true for all matrices Γ, Λ:*

$$\Gamma(I - \Lambda\Gamma)^{-1}\Lambda + I = (I - \Gamma\Lambda)^{-1}.$$

2.7 The Case of an Incomplete Market

In this section, stochastic volatility is now considered. Clearly, the results obtained in Sect. 2.6 could apply in this context whenever the market were assumed to be **completed** via an option-based hedging strategy. However, it is well-known that such an assumption is equivalent to a very demanding hypothesis on the realization of the options dynamics and their associated risk premia, and it may be more realistic to assume that the market remains incomplete, and then, study a hedging strategy based on the underlying asset only. As we shall see below, such a strategy leads to more involved pricing and hedging equations.

Let then the observed price process be a solution to the following set of SDE's

$$dS_t = S_t(\sigma_t dW_t^1 + \gamma\lambda d\delta_t + \mu_t dt) \tag{2.58}$$

$$d\sigma_t = \nu_t dt + \Sigma_t dW_t^2 \tag{2.59}$$

where (W^1, W^2) is a two-dimensional Wiener process under $\mathscr{P}$ with correlation ρ:

$$d < W^1, W^2 >_t = \rho dt,$$

and the processes μ_t, ν_t and Σ_t are actually functions of the state variables S, σ.

Consider again a Markovian framework, thereby looking for the value process V and the optimal strategy δ as smooth functions of the state variables

$$\delta_t = \delta(S_t, \sigma_t, t)$$
$$V_t = V(S_t, \sigma_t, t).$$

Then, the dynamics of the observed price becomes

$$dS_t = \frac{S_t}{1 - \gamma\lambda S_t \frac{\partial\delta}{\partial S}} \left(\sigma_t dW_t^1 + \gamma\lambda \frac{\partial\delta}{\partial\sigma} d\sigma_t + dQ_t \right), \tag{2.60}$$

the orthogonality condition reads

$$\left(\frac{\partial V}{\partial S} - \delta \right) d < S, \bar{S} >_t + \frac{\partial V}{\partial \sigma} d < \sigma, \bar{S} >_t = 0 \tag{2.61}$$

and the pricing equation for the value function V is

$$\frac{\partial V}{\partial t} + \frac{1}{2}\left(\frac{\partial^2 V}{\partial S^2} - \gamma\lambda S_t \left(\frac{\partial\delta}{\partial S} \right)^2 \right) \frac{d < S, S >_t}{dt} + \frac{1}{2}\left(\frac{\partial^2 V}{\partial \sigma^2} - \gamma\lambda S_t \left(\frac{\partial\delta}{\partial \sigma} \right)^2 \right) \frac{d < \sigma, \sigma >_t}{dt} +$$

$$+ \left(\frac{\partial^2 V}{\partial\sigma\partial S} - \gamma\lambda S_t \frac{\partial\delta}{\partial\sigma}\frac{\partial\delta}{\partial S} \right) \frac{d < S, \sigma >_t}{dt} + \mathscr{L}_1 V = 0, \tag{2.62}$$

where $\mathscr{L}_1$ is a first-order partial differential operator.

Equations (2.61) and (2.62) are quite complicated. In the next paragraph, we focus on a particular case that allows one to fully assess their well-posedness.

2.7.1 The Case $\gamma = 1$, $\rho = 0$

When $\gamma = 1$, the martingale component of the supply price does not depend on the strategy anymore. As a matter of fact, the supply price dynamics is given by

$$d\bar{S}_t = dS_t + S_t\left((1-2\gamma)\lambda d\delta_t + \frac{1}{2}\mu d < \delta, \delta >_t\right),$$

see (2.27), and therefore, using (2.58), there holds that

$$d\bar{S}_t = S_t(\sigma_t dW_t^1 + \lambda(1-\gamma)d\delta_t + dR_t) \equiv S_t(\sigma_t dW_t^1 + dR_t), \tag{2.63}$$

where R is a process of bounded variation. If, in addition, the Wiener processes for the asset and the volatility are supposed to be uncorrelated: $\rho = 0$, the tedious computations leading to the optimal hedge and value function simplify, and one can study in full generality the well-posedness of the pricing and hedging equations (2.61) and (2.62).

First and foremost, the orthogonality condition (2.61) simply reads in this case

$$\delta = \frac{\partial V}{\partial S}, \tag{2.64}$$

exactly as in the complete market case. This is a standard result in local-risk minimization with stochastic volatility when there is no correlation.

As for the pricing equation (2.62), one first works out using (2.64) the various brackets in (2.62) and finds that

$$\frac{d < S, S >_t}{dt} = \left(1 - \lambda S_t \frac{\partial^2 V}{\partial S^2}\right)^{-2} (\sigma_t^2 S_t^2 + \lambda^2 S_t^2 \left(\frac{\partial^2 V}{\partial S \partial \sigma})^2 \Sigma_t^2\right), \tag{2.65}$$

$$\frac{d < \sigma, \sigma >_t}{dt} = \Sigma^2 \tag{2.66}$$

and

$$\frac{d < S, \sigma >_t}{dt} = \left(1 - \lambda S_t \frac{\partial^2 V}{\partial S^2}\right)^{-1} \lambda S_t \Sigma_t^2 \frac{\partial^2 V}{\partial S \partial \sigma}. \tag{2.67}$$

Plugging these expressions in (2.62) yields the pricing equation for V

$$\frac{\partial V}{\partial t}+\frac{1}{2}\frac{\partial^2 V}{\partial S^2}\left(1-\lambda S\left(\frac{\partial^2 V}{\partial S^2}\right)\right)^{-1}\left(\sigma_t^2 S^2+\lambda^2 S^2\left(\frac{\partial^2 V}{\partial S\partial\sigma}\right)^2\Sigma_t^2\right)+\frac{1}{2}\left(\frac{\partial^2 V}{\partial\sigma^2}-\lambda S\left(\frac{\partial^2 V}{\partial S\partial\sigma}\right)^2\right)\Sigma^2+$$

$$\lambda S\Sigma_t^2\left(\frac{\partial^2 V}{\partial S\partial\sigma}\right)^2+\mathscr{L}_1 V=0, \tag{2.68}$$

or, after a few final rearrangements,

$$\frac{\partial V}{\partial t}+\frac{\sigma_t^2 S^2}{2(1-\lambda S(\frac{\partial^2 V}{\partial S^2}))}\frac{\partial^2 V}{\partial S^2}+\frac{1}{2}\frac{\partial^2 V}{\partial\sigma^2}\Sigma^2+\frac{1}{2}\frac{\lambda S\Sigma^2}{(1-\lambda S(\frac{\partial^2 V}{\partial S^2}))}\left(\frac{\partial^2 V}{\partial\sigma\partial S}\right)^2+\mathscr{L}_1 V=0. \tag{2.69}$$

The main result of this section is the

Proposition 2.3 *Equation* (2.69) *is of parabolic type.*

Proof One has to study the monotocity of the operator

$$\mathscr{L}: V\to\mathscr{L}(V)\equiv\frac{\sigma_t^2 S^2}{2(1-\lambda S(\frac{\partial^2 V}{\partial S^2}))}\frac{\partial^2 V}{\partial S^2}+\frac{1}{2}\frac{\partial^2 V}{\partial\sigma^2}\Sigma^2+\frac{1}{2}\frac{\lambda S\Sigma^2}{(1-\lambda S(\frac{\partial^2 V}{\partial S^2}))}\left(\frac{\partial^2 V}{\partial\sigma\partial S}\right)^2. \tag{2.70}$$

Introducing the classical notations

$$p\equiv\begin{pmatrix}p_{11} & p_{12}\\ p_{21} & p_{22}\end{pmatrix} \tag{2.71}$$

with $p_{11}=\frac{\partial^2 V}{\partial S^2}$, $p_{12}=p_{21}=\frac{\partial^2 V}{\partial S\partial\sigma}$ and $p_{22}=\frac{\partial^2 V}{\partial\sigma^2}$ and defining

$$\mathbf{L}(S,\mathbf{p})\equiv\frac{\sigma_t^2 S^2 p_{11}}{(1-\lambda S p_{11})}+\Sigma^2 p_{22}+\frac{\lambda S\Sigma^2}{(1-\lambda S p_{11})}p_{12}^2, \tag{2.72}$$

one is led to study the positivity of the 2×2 matrix

$$\begin{pmatrix}\frac{\partial\mathbf{L}}{\partial p_{11}} & \frac{1}{2}\frac{\partial\mathbf{L}}{\partial p_{12}}\\ \frac{1}{2}\frac{\partial\mathbf{L}}{\partial p_{12}} & \frac{\partial\mathbf{L}}{\partial p_{22}}\end{pmatrix}. \tag{2.73}$$

Setting $F(p_{11})=\frac{\sigma^2 S^2 p_{11}}{1-\lambda S p_{11}}$ and $D(p_{11})=1-\lambda S p_{11}$, one needs to show that the matrix $\mathbf{H}(\mathbf{p})$

$$\begin{pmatrix}F'(p_{11})+(\lambda S\Sigma)^2\frac{p_{12}^2}{D^2} & \lambda S\Sigma^2\frac{p_{12}}{D}\\ \lambda S\Sigma^2\frac{p_{12}}{D} & \Sigma^2\end{pmatrix} \tag{2.74}$$

is positive. This result is trivially shown to be true by computing the trace and determinant of $\mathbf{H}(\mathbf{p})$:

$$Tr(\mathbf{H}(\mathbf{p})) = F'(p_{11}) + \Sigma^2 + (\lambda S \Sigma)^2 \frac{p_{12}^2}{D^2} \tag{2.75}$$

and

$$Det(\mathbf{H}(\mathbf{p})) = \Sigma^2 F'(p_{11}) \tag{2.76}$$

and using the fact that F is a monotonically increasing function.

This ends the proof of Proposition 2.3.

As a final remark, we point out that the condition on the payoff for (2.69) to have a global, smooth solution, is exactly the same as in the one-dimensional case: stochastic volatility does not impose further constraints, except the now imperfect replication strategy.

2.8 Concluding Remarks

In this work, we model the effect of liquidity costs and market impact on the pricing and hedging of derivatives, using a static order book description and introducing a numerical parameter measuring the level of asymptotic market impact. In the complete market case, a structural result characterizing the well-posedness of the strategy-dependent diffusion is proven. Extensions to incomplete markets and non-linear hedging strategies are also considered.

We conclude with a discussion of the two conditions that play a fundamental role in our results.

2.8.1 *The Condition* $\gamma \in \left[\frac{2}{3}, 1\right]$

Of interest is the interpretation of the condition on the resilience parameter: $\frac{2}{3} \leqslant \gamma \leqslant 1$.

The case $\gamma > 1$ is rather trivial to understand, as one can easily see that it leads to arbitrage by a simple round-trip trade. The case $\gamma < \frac{2}{3}$ is not so simple. The loss of monotonicity of the function $F(p) = \frac{p(1+(1-2\gamma)p)}{(1-\gamma p)^2}$ for $\gamma < \frac{2}{3}$ yields the existence of p_1, p_2 such that $p_1 < p_2$ but $F(p_1) > F(p_2)$, which will lead to an inconsistency in the perfectly replicating strategies, as we now show.

Recall that the price of the replicating strategy solves the equation

$$\frac{\partial V}{\partial t} + \frac{1}{2}\sigma^2 S F\left(S\frac{\partial^2 V}{\partial S^2}\right) = 0, \tag{2.77}$$

and assume that there exists $p \in \mathbb{R}$ with $F'(p) < 0$. One can then find two values $p_1 < p_2$ such that $F(p_1) > F(p_2)$. Consider now two contingent claims Φ_1, Φ_2 satisfying $S\frac{\partial^2 \Phi_i}{\partial S^2} \equiv p_i, i = 1, 2$, together with $\frac{\partial \Phi}{\partial S}(S_0) = 0$, $\Phi_i(S_0) = 0$ for some given $S_0 > 0$. Under these assumptions, $\Phi_2(S) \geq \Phi_1(S)$ for all S. Then, there exist explicit solutions $V_i(t, S)$ to (2.77) with terminal conditions Φ_i, $i = 1, 2$, given simply by translations in time of the terminal payoff:

$$V_i(t, S) = \Phi_i(S) + (T - t)\frac{\sigma^2}{2} S F(p_i). \tag{2.78}$$

Consider the following strategy: sell the terminal payoff Φ_1 at price $V_1(0, S_0)$, without hedging, and hedge Φ_2 following the replicating strategy given by (2.77).

The final wealth of such a strategy is given by

$$\text{Wealth}(T) = \underbrace{(\Phi_2(S_T) - V_2(0, S_0))}_{\text{hedge strategy}} + \underbrace{(V_1(0, S_0) - \Phi_1(S_T))}_{\text{option sold}}. \tag{2.79}$$

Using (2.78), one obtains

$$\text{Wealth}(T) = T\frac{\sigma^2}{2} S_0(F(p_1) - F(p_2)) + (\Phi_2(S_T) - \Phi_2(S_0)) - (\Phi_1(S_T) - \Phi_1(S_0)), \tag{2.80}$$

which is always positive, given the conditions on Φ_1, Φ_2, and thereby generates what may be interpreted as an arbitrage opportunity.

Note that this arbitrage exists both for $\gamma > 1$ and $\gamma < 2/3$, since it just requires that F be locally decreasing. However, in the case $\gamma > 1$, round-trip trades generate money and the price dynamics create actual arbitrage opportunities, whereas in the case $\gamma < 2/3$, it is the option prices generated by exact replication strategies that lead to a potential arbitrage: in order to make a profit, one should find a counterparty willing to buy an option at its exact replication price.

It is clear that such a "counterexample" is not an arbitrage opportunity per se, as one has to find a counterparty to this contract—what this means is simply that the price of the perfect hedge is not the right price for the option.

2.8.2 The Condition $S\frac{\partial^2 V}{\partial S^2} < \frac{1}{\gamma\lambda}$

Another important question has been left aside so far: the behaviour of the solution to the pricing equation when the constraint is violated at maturity—after all, this is bound to be the case for a real-life contingent claim such as a call option! From a mathematical point of view, see the discussion in Loeper (2013), there is a solution which amounts to replace the pricing equation $\mathscr{P}(D)(V) = 0$ by

$\text{Max}(\mathscr{P}(D)(V), S\frac{\partial^2 V}{\partial S^2} - \frac{1}{\gamma\lambda}) = 0$, but of course, in this case, the perfect replication does not exist any longer—one should use a super-replicating strategy as introduced originally in Soner and Touzi (2000) exactly for this purpose.

References

F. Abergel and N. Millot. Non quadratic local risk-minimization for hedging contingent claims in incomplete markets. *SIAM Journal on Financial Mathematics*, 2(1):342–356, 2011.

R. Almgren. Option hedging with market impact. *Presentation at the Market Microstructure: Confronting Many Viewpoints conference, Paris*, 2012.

R. Almgren, C. Thum, E. Hauptmann, and H. Li. Direct estimation of equity market impact. *Risk*, 18:58–62, 2005.

N. Bershova and D. Rakhlin. The non-linear market impact of large trades: Evidence from buy-side order flow. *Quantitative finance*, 13:1759–1778, 2013.

U. Cetin, R. Jarrow, and P. Protter. Liquidity risk and arbitrage pricing theory. *Finance and Stochastics*, 8:311–341, 2004.

U. Çetin, H. M. Soner, and N. Touzi. Option hedging for small investors under liquidity costs. *Finance Stoch.*, 14(3):317–341, 2010.

J. D. Farmer, A. Gerig, F. Lillo, and H. Waelbroeck. How efficiency shapes market impact. *Quantitative finance*, 13:1743–1758, 2013.

R. Frey. Perfect option hedging for a large trader. *Finance and Stochastics*, 2(2):115–141, 1998.

R. Frey and A. Stremme. Market volatility and feedback effects from dynamic hedging. *Mathematical Finance*, 7(4):351–374, 1997.

D. Gilbarg and N. S. Trudinger. *Elliptic partial differential equations of second order*. Springer, 1998.

O. A. Ladyzhenskaya, V. A. Solonnikov, and N. N. Uraltseva. *Linear and quasilinear equations of parabolic type*. American Mathematical Society, 1968.

D. Lamberton, H. Pham, and M. Schweizer. Local risk-minimization under transaction costs. *Mathematics of Operations Research*, 23:585–612, 1997.

H. E. Leland. Option pricing and replication with transactions costs. *Journal of Finance*, 40(5):1283–1301, 1985.

H. Liu and J. M. Yong. Option pricing with an illiquid underlying asset market. *Journal of Economic Dynamics and Control*, 29:2125–2156, 2005.

G. Loeper. Option pricing with market impact and non-linear black and scholes pde's. arXiv:1301.6252, 2013.

N. Millot. *Hedging Contingent Claims by Convex Local Risk-Minimization*. PhD thesis, 2012.

N. Millot and F. Abergel. Non quadratic local risk-minimization for hedging contingent claims in the presence of transaction costs. *Available at SSRN 1881175*, 2011.

E. Platen and M. Schweizer. On feedback effects from hedging derivatives. *Mathematical Finance*, 8(1):67–84, 1998.

A. Roch. *Liquidity risk, volatility and financial bubbles*. PhD thesis, 2009.

P. J. Schönbucher and P. Wilmott. The feedback effect of hedging in illiquid markets. *SIAM J. Appl. Maths*, 61(1):232–272, 2000.

H. M. Soner and N. Touzi. Supperreplication under gamma constraints. *SIAM Journal on Control and Optimization*, 39(1):73–96, 2000.

L. Wang. On the regularity theory of fully nonlinear parabolic equations i. *Communications on Pure and Applied Mathematics*, 45:27–86, 1992a.

L. Wang. On the regularity theory of fully nonlinear parabolic equations ii. *Communications on Pure and Applied Mathematics*, 45:141–178, 1992b.

P. Weber and B. Rosenow. Order book approach to price impact. *Quantitative Finance*, 5(4):357–364, 2005.

Valeri I. Zakamouline. European option pricing and hedging with both fixed and proportional transaction costs. *Journal of Economic Dynamics and Control*, 30(1):1–25, 2006.

Chapter 3
Dynamic Portfolio Credit Risk and Large Deviations

Sandeep Juneja

Abstract We consider a multi-time period portfolio credit risk model. The default probabilities of each obligor in each time period depend upon common as well as firm specific factors. The time movement of these factors is modelled as a vector autoregressive process. The conditional default probabilities are modelled using a general representation that subsumes popular default intensity models, logit-based models as well as threshold based Gaussian copula models. We develop an asymptotic regime where the portfolio size increases to infinity. In this regime, we conduct large deviations analysis of the portfolio losses. Specifically, we observe that the associated large deviations rate function is a solution to a quadratic program with linear constraints. Importantly, this rate function is independent of the specific modelling structure of conditional default probabilities. This rate function may be useful in identifying and controlling the underlying factors that contribute to large losses, as well as in designing fast simulation techniques for efficiently measuring portfolio tail risk.

3.1 Introduction

Financial institutions such as banks have portfolio of assets comprising thousands of loans, defaultable bonds, credit sensitive instruments and other forms of credit exposures. Calculating portfolio loss distribution at a fixed time in future as well as its evolution as a function of time, is crucial to risk management: Of particular interest are computations of unexpected loss or tail risk in the portfolio. These values

---

CAFRAL (Centre for Advanced Financial Research and Learning), a research wing of Reserve Bank of India. This research was conducted primarily at CAFRAL.

S. Juneja (✉)
TIFR, Mumbai, India
e-mail: juneja@tifr.res.in

F. Abergel et al. (eds.), *Econophysics and Sociophysics: Recent Progress and Future Directions*, New Economic Windows,
DOI 10.1007/978-3-319-47705-3_3

are important inputs to the amount of capital an institution may be required to hold for regulatory purposes. There is also interest in how this capital requirement evolves over time.

In this short note, we develop a discrete time dynamic model that captures the stochastic evolution of default probabilities of different firms in the portfolio as a function of time. We then develop an asymptotic regime to facilitate analysis of tail distribution of losses. We conduct large deviations analysis of losses in this regime identifying the large deviations rate function associated with large losses. This tail analysis provides great deal of insight into how large losses evolve over time in a credit portfolio.

There is a vast literature on modelling credit risk and on modelling a portfolio of credit risk (see, e.g., Duffie and Singleton 2012; Merton 1974; Giesecke et al. 2011). Glasserman and Li (2005), Dembo et al. (2004), Glasserman et al. (2007), Bassamboo et al. (2008), Zhang et al. (2015) are some of the works that conduct large deviations analysis for large portfolio losses in a static single period setting.

Our contributions: As mentioned earlier, we model the evolution of the credit portfolio in discrete time. The conditional probabilities of default of surviving firms in any time period is modelled as a function of a linear combination of stochastic covariates. This subsumes logit function models for conditional probabilities, default intensity models (in discrete time) as well as threshold based Gaussian and related copula models (see Duffie et al. 2007; Duan et al. 2012; Duan and Fulop 2013; Chava and Jarrow 2004; Sirignano and Giesecke 2015 as examples where similar dependence on stochastic covariates is considered). We model the stochastic evolution of the stochastic covariates as a vector AR process, although the essential features of our analysis are valid more broadly.

As is a common modelling practice, we assume that these stochastic variates are multivariate Gaussian distributed, and can be classified as:

- *Systemic common covariates*: These capture macroeconomic features such as GDP growth rates, unemployment rates, inflation, etc.
- *Class specific covariates*: All loans in our portfolio belong to one of a fixed number of classes. These capture the common exposure to risk to obligors in the same industry, geographic region, etc.
- *Idiosyncratic variates*: This captures the idiosyncratic risk corresponding to each obligor.

We embed the portfolio risk problem in a sequence of problems indexed by the portfolio size n. We develop an asymptotic regime where the conditional default probabilities decrease as n increases. In this regime we identify the large deviations rate function of the probability of large losses at any given time in future. Our key contribution is to show that the key component to ascertaining this rate function is a solution to a quadratic program with linear constraints. Further we observe in specialized settings that the resultant quadratic program can be explicitly solved to give a simple expression for the large deviations rate function. Our other contribution is to highlight that in a fairly general framework, the underlying structure of how portfolio losses build up is independent of the specific model for conditional default probabilities—thus whether we use default intensity model, logit based model or a Gaussian Copula based model for default probabilities, to the first order (that is, on the large deviations scaling), the portfolio tail risk measurement is unaffected.

Our large deviations analysis may be useful in identifying and controlling parameters that govern the probability of large losses. It is also critical to development of fast simulation techniques for the associated rare large loss probabilities. Development of such techniques is part of our ongoing research and not pursued here. In this paper, we assume that each class has a single class specific covariate and these are independent of all other covariates. This is a reasonable assumption in practice and makes the analysis substantially simpler. As we discuss later in Sect. 3.3, relaxing this and many other assumptions, is part of our ongoing research that will appear separately.

Roadmap: In Sect. 3.2, we develop the mathematical framework including the asymptotic regime for our analysis. We end with a small conclusion and a discussion of our ongoing work in Sect. 3.3. Some of the technical details are kept in Appendix.

3.2 Mathematical Model

Consider a portfolio credit risk model comprising n obligors. These are divided into K classes $\{1, 2, \ldots, K\}$, $\mathscr{C}_j$ denotes the obligors in class j. As mentioned in the introduction, we model conditional default probabilities using structures that subsume discrete default intensity models considered in Duffie et al. (2007) as well as Duan et al. (2012), popular logit models (see, e.g., Chava and Jarrow 2004; Sirignano and Giesecke 2015), as well as threshold based Gaussian and related copula models (see, e.g., Glasserman and Li 2005; Glasserman et al. 2007; Bassamboo et al. 2008).

First consider the discrete default intensity model and suppose that time horizon of our analysis is a positive integer τ. We restrict ourselves to discrete default intensities taking the proportional-hazards form as in Duffie et al. (2007), Duan et al. (2012). Specifically, suppose that one period conditional default probability for a firm i in $\mathscr{C}_j$, at period $t \leq \tau$, is given by

$$p_{i,j,t} = 1 - \exp[-\exp(P_{i,j,t})],$$

where,

$$P_{i,j,t} = -\alpha_j + \beta^T \mathbf{F}_t + \gamma_j G_{j,t} + \varepsilon_{i,t},$$

where the above variables have the following structure:

- For $j \leq K, \alpha_j > 0$ and for $d \geq 1$, $\beta \in \Re^d$. $(\gamma_j, j \in K)$ are w.l.o.g. non-negative constants.
- Random variables $\varepsilon = (\varepsilon_{i,t} : i \leq n, t \leq \tau)$ are assumed to be i.i.d. (independent, identically distributed) with standard Gaussian distribution with mean zero and variance one.

- $(\mathbf{F}_t \in \Re^d : t = 0, \ldots, \tau)$ denote the common factors that affect default probabilities of each obligor. To keep the analysis simple we assume that $(\mathbf{F}_t : t = 0, \ldots, \tau)$ follows the following VAR(1) process.

$$\mathbf{F_{t+1}} = \mathbf{A}\mathbf{F}_t + \tilde{\mathbf{E}}_{t+1}$$

where $\mathbf{A} \in \Re^{d\times d}$ and $\tilde{\mathbf{E}} = (\tilde{\mathbf{E}}_t : t = 1, \ldots, \tau)$ is a sequence of i.i.d. random vectors, assumed to be Multi-variate Gaussian with mean 0 and positive definite variance covariance matrix Σ. Further let $\mathbf{B}$ be a matrix such that $\mathbf{B}\mathbf{B}^T = \Sigma$. Then, we can model

$$\mathbf{F_{t+1}} = \mathbf{A}\mathbf{F}_t + \mathbf{B}\mathbf{E}_{t+1}$$

where, each $\mathbf{E}_t = (E_{t,j} : j \leq d)$ is a vector of independent mean zero, variance one, Gaussian random variables. Then, it follows that for $t \geq 1$,

$$\mathbf{F_t} = \mathbf{A}^t\mathbf{F}_0 + \sum_{i=1}^{t} \mathbf{A}^{t-i}\mathbf{B}\mathbf{E}_i.$$

- The random variables $(G_{j,t} : j \leq K, t \leq \tau)$ capture the residual class risk (once the risk due to the common factors is accounted for by $\tilde{\mathbf{E}}$). These are assumed to be independent of $\tilde{\mathbf{E}}$ as well as ε. Further, we assume that they follow a simple autoregressive structure

$$G_{j,t} = \eta_j G_{j,t-1} + \Lambda_{j,t}$$

where $(\Lambda_{j,t} : j \leq K, t \leq \tau)$ are assumed to be i.i.d., mean zero, variance one, standard Gaussian distributed.
- It follows that

$$G_{j,t} = \eta_j^t G_{j,0} + \sum_{i=1}^{t} \eta_j^{t-i} \Lambda_{j,i}. \tag{3.1}$$

To keep the analysis notationally simple, we assume that exposure e_i of each obligor $i \in \mathscr{C}_j$ equals ex_j. This denotes the amount lost if an obligor in $\mathscr{C}_j$ defaults net of recoveries made on the loan.

An analogous logit structure for conditional probabilities corresponds to setting

$$p_{i,j,t} = \frac{\exp(P_{i,j,t})}{1 + \exp(P_{i,j,t})}.$$

In the remainder of the paper, we assume that

$$p_{i,j,t} = F(P_{i,j,t})$$

where $F : \Re \to [0, 1]$ is a distribution function that we assume is strictly increasing. Thus, $F(-\infty) = 0$ and $F(\infty) = 1$. In the setting of Logit function

$$F(x) = \frac{e^{\theta x}}{1 + e^{\theta x}} \tag{3.2}$$

and for default intensity function

$$F(x) = 1 - \exp(-e^{\theta x}) \tag{3.3}$$

for $\theta > 0$.

Another interesting setting to consider is the J.P. Morgan's threshold based Gaussian Copula models extensively studied in literature, see, e.g., Glasserman and Li (2005) and Glasserman et al. (2007). Adapting this approach to our setting, an obligor i in class j that has survived till time $t - 1$, defaults at time t if

$$\beta^T \mathbf{F}_t + \gamma_j G_{j,t} + \varepsilon_{i,t} > \alpha_j$$

for large α_j. These models are studied in literature for a single time period, but can be generalized for multiple time periods by having a model for time evolution of common and class specific factors, as we consider in this paper.

One way to concretely fit this to our outlined framework, express

$$\varepsilon_{i,t} = \frac{\varepsilon_{i,t}(1) + \varepsilon_{i,t}(2)}{\sqrt{2}}$$

where $\varepsilon_{i,t}(1)$ and $\varepsilon_{i,t}(2)$ are independent Gaussian mean zero, variance one random variables. Then, set

$$P_{i,j,t} = -\alpha_j + \beta^T \mathbf{F}_t + \gamma_j G_{j,t} + \frac{1}{\sqrt{2}} \varepsilon_{i,t}(1)$$

to get

$$p_{i,j,t} = F(P_{i,j,t}) = \bar{\Phi}(-P_{i,j,t}) \tag{3.4}$$

where $\bar{\Phi}(\cdot)$ denotes the tail distribution function of a mean zero, variance half, Gaussian random variable (here $F(x) = \bar{\Phi}(-x)$).

3.2.1 Probability of Large Losses

In this note, our interest is in developing large deviations asymptotic for the probability of large losses in the portfolio by any specified time τ. It may be useful to spell out a Monte Carlo algorithm to estimate the probability that portfolio losses L by time τ exceed a large threshold u.

Monte Carlo Algorithm: Suppose that the current time is zero and our interest is in generating via simulation independent samples of portfolio losses by time τ. We assume that $\mathbf{F}_0$ and $(G_j(0) : j \leq K)$ are available to us.

In the algorithm below, let $\mathscr{S}_t$ denote the surviving, non-defaulted obligors at (just after) time t and $\mathscr{L}_t$ denote the losses incurred at time t. $\mathscr{S}_0$ denotes all the obligors. The algorithm then proceeds as follows

1. Set time $t = 1$.
2. While $t \leq \tau$,
   a. Generate independent samples of $(\varepsilon_{i,t} : i \in \mathscr{S}_{t-1})$, $\mathbf{E}_t$ and $(\Lambda_{j,t} : j \leq K)$ and compute $p_{i,j,t}$ for each $(i \in \mathscr{S}_{t-1}, j \leq K)$.
   b. Generate independent uniform numbers $(U_{i,t} : i \in \mathscr{S}_{t-1})$. Obligor $i \in \mathscr{S}_{t-1}$ defaults at time t if $U_{i,t} \leq p_{i,j,t}$. Recall that obligor $i \in \mathscr{C}_j$ causes loss e_j if it defaults. Compute $\mathscr{S}_t$ as well as $\mathscr{L}_t$.
3. A sample of total loss by time T is obtained as $L = \sum_{t=1}^{\tau} \mathscr{L}_t$.
4. Set $I(L > u)$ to one if the loss L exceeds u and zero otherwise. Sample average of independent samples of $I(L > u)$ then provides an unbiased and consistent estimator of $P(L > u)$.

As mentioned in the introduction, we analyze the probability of large losses in an asymptotic regime that we develop in Sect. 3.2.2.

3.2.2 Asymptotic Regime

Let $(\mathscr{P}_n : n \geq 1)$ denote a sequence of portfolios. $\mathscr{P}_n$ denotes a portfolio with n obligors. As before the size of class $\mathscr{C}_k$ in $\mathscr{P}_n$ equals $c_k n$ so that $\sum_{k=1}^{K} c_k = 1$. To avoid unnecessary notational clutter we assume that $c_k n$ is an integer for each k and n.

In $\mathscr{P}_n$, for each n, the conditional probability of default $p_{i,j,t}(n)$ at time t for obligor $i \in \mathscr{C}_j$ that has not defaulted by time $t - 1$ is denoted by $F(P_{i,j,t}(n))$, where

$$P_{i,j,t}(n) = -\alpha_j m_n + \tilde{m}_n \beta^T \mathbf{F}_t + \tilde{m}_n \gamma_j G_{j,t} + \tilde{m}_n \varepsilon_{i,t}$$

for each n, i and t. Here, m_n and $\tilde{m}_n$ are positive sequences increasing with n. The sequence of random vectors $(\mathbf{F}_t : t \leq \tau)$ and $(G_{j,t} : j \leq K, t \leq \tau)$ evolve as specified in the previous section and notations $(\mathbf{E}_t : t \leq \tau)$ and $(\Lambda_{j,t} : j \leq K, t \leq \tau)$ remain unchanged. For notations $(\mathscr{S}_t, \mathscr{L}_t : t \leq \tau)$, $(U_{i,t} : i \leq n, t \leq \tau)$ we simply suppress dependence on n for presentation simplicity. Here, $(U_{i,t} : i \leq n, t \leq \tau)$ are used to facilitate Monte Carlo interpretation of defaults.

The following assumption is needed.

Assumption 1

$$\limsup_{n\to\infty} r_n = \frac{m_n}{\tilde{m}_n} = \infty. \tag{3.5}$$

Remark 3.1 There is a great deal of flexibility in selecting $\{m_n\}$ and $\{\tilde{m}_n\}$ allowing us to model various regimes of default probabilities. When r_n increases to infinity at a fast rate, the portfolio comprises obligors with small default probabilities. When it goes to infinity at a slow rate, the portfolio comprises obligors with relatively higher default probabilities.

Let $\tilde{A}_{i,t}$ denote the event that obligor i defaults at time t in $\mathscr{P}_n$, i.e., $i \in \mathscr{S}_{t-1}$ and $U_{i,t} \le p_{i,j,t}(n)$. Then,

$$A_{i,t} = \cup_{s=1}^{t} \tilde{A}_{i,s}$$

denotes the event that obligor i defaults by time t.

The aim of this short note is to develop the large deviations asymptotics for the probabilities

$$P\left(\sum_{i=1}^{n} e_i I(A_{i,\tau}) > na\right)$$

as $n \to \infty$.

Note that obligor $i \in \mathscr{S}_{t-1} \cap \mathscr{C}_j$ defaults at time t if

$$U_{i,t} \le F(P_{i,j,t}(n)).$$

Equivalently, if

$$P_{i,j,t}(n) \ge F^{-1}(U_{i,t}).$$

This in turn corresponds to

$$-m_n\alpha_j + \tilde{m}_n\beta^T(\mathbf{A}^t\mathbf{F}_0 + \textstyle\sum_{i=1}^{t} \mathbf{A}^{t-i}\mathbf{B}\mathbf{E}_i) +$$
$$\tilde{m}_n(\eta^t\gamma_j G_{j,0} + \gamma_j \textstyle\sum_{i=1}^{t} \eta^{t-i}\Lambda_{j,i}) + \tilde{m}_n\varepsilon_{i,t} \ge F^{-1}(U_{i,t}). \tag{3.6}$$

Let $H_t = \beta^T(\sum_{j=1}^{t} \mathbf{A}^{t-j}\mathbf{B}\mathbf{E}_j)$. For each i, j, let $\mathbf{h}_j = (h_{j,k} : 1 \le k \le d)$ be defined by

$$\mathbf{h}_j = \beta^T\mathbf{A}^j\mathbf{B}.$$

Recall that $\mathbf{E}_t = (E_{t,k} : k \le d)$ is a vector of independent mean zero variance 1, Gaussian random variables. Thus, we may re-express

$$H_t = \sum_{j=1}^{t}\sum_{k=1}^{d} h_{t-j,k}E_{j,k}.$$

Then H_t is a mean zero Gaussian random variable with variance

$$v(H_t) = \beta^T \left(\sum_{j=1}^{t} \mathbf{A}^{t-j} \Sigma (\mathbf{A}^{t-j})^T \right) \beta.$$

Let $Y_{j,t} = \gamma_j \sum_{k=1}^{t} \eta^{t-k} \Lambda_{j,k}$ and for $i \in \mathscr{C}_j$,

$$Z_{i,t}(n) = \varepsilon_{i,t} - \tilde{m}_n^{-1} F^{-1}(U_{i,t}) + \beta^T \mathbf{A}^t \mathbf{F}_0 + \eta^t \gamma_j G_{j,0}.$$

Then, $\tilde{A}_{i,t}$ occurs if $i \in \mathscr{S}_{t-1} \cap \mathscr{C}_j$ and

$$H_t + Y_{j,t} \geq r_n \alpha_j - Z_{i,t}(n).$$

Below we put a mild restriction on m_n, $\tilde{m}_n$, tail distribution of each $\varepsilon_{i,t}$, and the functional form of F:

Assumption 2 There exists a non-negative, non-decreasing function g such that $g(x) \to \infty$ as $x \to \infty$, and

$$\limsup_n \sup_{t \leq \tau, j \leq K, i \in \mathscr{C}_j} P(Z_{i,t}(n) \geq x) \leq e^{-g(x)}.$$

Further, there exists a $\delta \in (0, 1)$ such that

$$\liminf_{n \to \infty} \frac{g(r_n^\delta) n}{r_n^2} = +\infty. \tag{3.7}$$

Remark 3.2 Since, for fixed $\mathbf{F}_0$ and $G_{j,0}$, the term $\beta^T \mathbf{A}^t \mathbf{F}_0 + \eta^t \gamma_j G_{j,0}$ can be uniformly bounded by a constant, call it c, and

$$P(\varepsilon_{i,t} - \tilde{m}_n^{-1} F^{-1}(U_{i,t}) \geq x - c) \leq P(\varepsilon_{i,t} \geq (x-c)/2) + P(-\tilde{m}_n^{-1} F^{-1}(U_{i,t}) \geq (x-c)/2),$$

in Assumption 2, the key restriction is imposed by the tail distribution of $-\tilde{m}_n^{-1} F^{-1}(U_{i,t})$ and we look for a function g and $\delta \in (0, 1)$ such that

$$P(-\tilde{m}_n^{-1} F^{-1}(U_{i,t}) \geq x) \leq e^{-g(x)} \tag{3.8}$$

for all sufficiently large n, and (3.7) holds. Equation (3.8) is equivalent to finding g so that

$$\log \left(\frac{1}{F(-\tilde{m}_n x)} \right) \geq g(x), \tag{3.9}$$

for all sufficiently large n. Consider first the case of F in (3.2) as well as (3.3). In that case, the LHS is similar to

$$\theta \tilde{m}_n x$$

for large $\tilde{m}_n x$, and condition (3.7) holds if $\tilde{m}_n$, r_n and $\delta \in (0, 1)$ are selected so that

$$\frac{\tilde{m}_n r_n^\delta n}{r_n^2} \to \infty.$$

This is achieved, for instance, if for $\kappa \in (0, 1)$, $r_n = n^\kappa$, and

$$2 - 1/\kappa < \delta < 1,$$

for arbitrarily increasing $\{\tilde{m}_n\}$.

Now consider F in (3.4) where $F(x) = \bar{\Phi}(-x)$. Then, LHS of (3.9) is similar to

$$\tilde{m}_n^2 x^2$$

for large $\tilde{m}_n x$, and condition (3.7) holds if $\tilde{m}_n$, r_n and δ are selected so that

$$\frac{\tilde{m}_n r_n^\delta n}{r_n^2} \to \infty.$$

This is achieved, for instance, if for $\kappa > 0$, $r_n = n^\kappa$, and

$$1 - 1/(2\kappa) < \delta < 1,$$

for arbitrarily increasing $\{\tilde{m}_n\}$.

Let

$$N_j(t) = \sum_{i \in \mathscr{C}_j} I(\tilde{A}_{i,t})$$

denote the number of defaults for class j at time t for each $j \le K$ and $t \le \tau$.

Let

$$\mathscr{N} = \{\frac{N_1(\tau)}{n} \ge a_\tau\},$$

where $a_\tau \in (0, c_1)$.

In Theorem 3.1 below we argue that on the large deviations scaling, the probability of default of any fraction of total customers in a single class at a particular time equals the probability that the complete class defaults at that time. It also highlights the fact that in the single class setting, in our regime, large losses are much more likely to occur later rather than earlier. This then provides clean insights into how large losses happen in the proposed regime.

Theorem 3.1 *Under Assumptions* 1 *and* 2,

$$\lim_{n\to\infty} \frac{1}{r_n^2} \log P(\mathcal{N}) = -q^*(\tau),$$

where $q^*(t)$ *is the optimal value of the quadratic program*

$$\min \sum_{k=1}^{t} \sum_{p=1}^{d} e_{k,p}^2 + \sum_{k=1}^{t} l_k^2,$$

subject to,

$$\sum_{k=1}^{t} \sum_{p=1}^{d} h_{t-k,p} e_{k,p} + \gamma_j \sum_{k=1}^{t} \eta_1^{t-k} l_k \geq \alpha_1,$$

and, for $1 \leq \tilde{t} \leq t-1$,

$$\sum_{k=1}^{\tilde{t}} \sum_{p=1}^{d} h_{\tilde{t}-k,p} e_{k,p} + \gamma_j \sum_{k=1}^{\tilde{t}} \eta_1^{\tilde{t}-k} l_k \leq \alpha_1.$$

Further, $q^*(t)$ *equals*

$$\frac{\alpha_1^2}{\sum_{k=1}^{t} \sum_{p=1}^{d} h_{t-k,p}^2 + \gamma_j^2 \sum_{k=1}^{t} \eta_1^{2(t-k)}}. \tag{3.10}$$

Note that it strictly reduces with t.

Remark 3.3 In Theorem 3.1, its important to note that $q^*(\tau)$ is independent of the values $a_\tau \in (0, c_1)$.

Some notation, and Lemma 3.1 are needed for proving Theorem 3.1. For each $j \leq K$, let

$$\mathcal{H}_{j,t} = \{H_t + Y_{j,t} \geq r_n \alpha_j + r_n^\delta\}$$

and

$$\tilde{\mathcal{H}}_{j,t} = \{H_t + Y_{j,t} \leq r_n \alpha_j - r_n^\delta\}.$$

Let,

$$\mathcal{H}_j^t = \left(\mathcal{H}_{j,t} \cap (\cap_{\tilde{t}=1}^{t-1} \tilde{\mathcal{H}}_{j,\tilde{t}})\right).$$

Lemma 3.1

$$\lim_{n\to\infty} \frac{1}{r_n^2} \log P(\mathcal{H}_1^t) = -q^*(t),$$

where $q^*(t)$ *is defined in* (3.10).

3.2.3 Key Result

Recall that our interest is in developing large deviations asymptotics for $P(\sum_{i=1}^{n} e_i I(A_{i,\tau}) > na)$.

Let $\mathbf{b}$ denote a sub-collection of indices of $\{1, 2, \ldots, K\}$ such that $\sum_{i\in\mathbf{b}} ex_j c_j > a$, while this is not true for any subset of $\mathbf{b}$. We call such a set $\mathbf{b}$ a minimal set, and we let $\mathscr{B}$ denote a collection of all such minimal sets (similar definitions arise in Glasserman et al. 2007). Consider the question that losses from the portfolio exceed na when we count losses only from classes indexed by $\mathbf{b}$. Our analysis from Theorem 3.1 can be repeated with minor adjustments to conclude that the large deviations rate for this is the smallest of all solutions to the quadratic programs of the form described below.

For $\mathbf{t} = (t_j, j \in \mathbf{b})$ such that each $t_j \le \tau$. Set $t_{\max} = \max_{j\in\mathbf{b}} t_j$ and let $q^*(\mathbf{t}, \mathbf{b})$ be the solution to quadratic program below (call it **O2**),

$$\min \sum_{k=1}^{t_{\max}} \sum_{p=1}^{d} e_{k,p}^2 + \sum_{j\in\mathbf{b}} \sum_{k=1}^{t_j} l_{j,k}^2$$

subject to, for all $j \in \mathbf{b}$,

$$\sum_{k=1}^{t_j} \sum_{p=1}^{d} h_{t_j-k,p} e_{k,p} + \gamma_j \sum_{k=1}^{t_j} \eta_j^{t_j-k} l_{j,k} \ge \alpha_j,$$

and, for $1 \le \tilde{t} \le t_j - 1$,

$$\sum_{k=1}^{\tilde{t}} \sum_{p=1}^{d} h_{\tilde{t}-k,p} e_{k,p} + \gamma_j \sum_{k=1}^{\tilde{t}} \eta_j^{\tilde{t}-k} l_k \le \alpha_j.$$

Set

$$\tilde{q}(\tau, \mathbf{b}) = \min_{\mathbf{t}: j\in\mathbf{b}, t_j \le \tau} q^*(\mathbf{t}, \mathbf{b}).$$

It is easy to see that there exists an optimal $\mathbf{t}^*$ such that $\tilde{q}(\tau, \mathbf{b}) = q^*(\mathbf{t}^*, \mathbf{b})$ with the property that the respective constraints for each $\tilde{t} < t_j^*$ are not tight. Whenever, there exists $\tilde{t} < t_j$ such that constraint corresponding to $\tilde{t}$ is tight, a better rate function value is achieved by setting such a $t_j = \tilde{t}$. This then helps complete the proof of the large deviations result. It is also then easy to see that the most likely way for $\{\sum_{i=1}^{n} e_i I(A_{i,t}) > na\}$ to happen is that all obligors belonging to class $\mathbf{b}$ default by time t, where $\mathbf{b}$ is selected as the most likely amongst all the classes in $\mathscr{B}$. In other words,

$$\lim_{n\to\infty} \frac{1}{r_n^2} \log P(\sum_{i=1}^{n} e_i I(A_{i,\tau}) > na) = -\min_{\mathbf{b}\in\mathscr{B}} \tilde{q}(\tau, \mathbf{b}).$$

The proof again is straightforward and relies on the following simple lemma (see, e.g., Dembo and Zeitouni 2009, Lemma 1.2.15).

Lemma 3.2 *Let N be a fixed integer. Then, for every* $a_\varepsilon^i \geq 0$,

$$\limsup_{\varepsilon \to \infty} \varepsilon \log \left(\sum_{i=1}^{N} a_\varepsilon^i \right) = \max_{i \leq N} \limsup_{\varepsilon \to \infty} \varepsilon \log a_\varepsilon^i.$$

In particular, the lim sup *above can be replaced by* lim *above if* $\max_{i \leq N} \lim_{\varepsilon \to \infty} \varepsilon \log a_\varepsilon^i$ *exists.*

3.2.4 Single Period Setting

In practice, one is often interested in solving for portfolio credit risk in a single period, that is, $\tau = 1$. In that case, it is easy to arrive at a simple algorithm to determine $q^*(1, \mathbf{b})$ and the associated values of the variables.

Note that the optimization problem **O2** reduces to

$$\min \sum_{p=1}^{d} e_{1,p}^2 + \sum_{j \in \mathbf{b}} l_{j,1}^2,$$

subject to, for all $j \in \mathbf{b}$,

$$\sum_{p=1}^{d} h_{0,p} e_{1,p} + \gamma_j l_{j,1} \geq \alpha_j.$$

Call this problem **O3**. The following remark is useful in solving **O3**.

Remark 3.4 It is easy to see that for the optimization problem—minimize $\sum_{k=1}^n x_k^2$ subject to

$$\sum_{k=1}^{n} a_k x_k \geq b, \tag{3.11}$$

the solution for each k is

$$x_k^* = b \frac{a_k}{\sum_{j=1}^n a_j^2}$$

and (3.11) is tight. The optimal objective function value is

$$\frac{b^2}{\sum_k a_k^2}.$$

To simplify the notation, suppose that $\mathbf{b} = \{1, 2, \ldots, k\}$ and that $\alpha_1 \geq \alpha_2 \geq \alpha_k > 0$. In view of Remark 3.4, solving **O3** can be reduced to solving the quadratic program

$$\min \quad cx^2 + \sum_{j \leq k} c_j y_j^2$$

subject to

$$x + y_j \geq \alpha_j \tag{3.12}$$

for all $j \leq k$, where $c = 1/(\sum_{p=1}^{d} h_{0,p}^2)$ and $c_j = 1/\gamma_j^2$ for each j.

It is easily seen using the first order condition that there exists a $1 \leq j^* \leq k$ such that under the unique optimal solution, constraints (3.12) hold as equalities for $1 \leq j \leq j^*$, that optimal x^* equals

$$\frac{\sum_{j \leq j^*} c_j \alpha_j}{c + \sum_{j \leq j^*} c_j}$$

and this is $\leq \alpha_{j^*}$. Then, $y_j = \alpha_j - x^*$ for $j \leq j^*$, and $y_j = 0$ otherwise.

Further, the optimal objective function equals,

$$(c + \sum_{j \leq j^*} c_j) \left(\frac{\sum_{j \leq j^*} c_j \alpha_j^2}{c + \sum_{j \leq j^*} c_j} - \left(\frac{\sum_{j \leq j^*} c_j \alpha_j}{c + \sum_{j \leq j^*} c_j} \right)^2 \right).$$

The algorithm below to ascertain j^* is straightforward.

Algorithm

1. If

$$\frac{\sum_{j \leq k} c_j \alpha_j}{c + \sum_{j \leq k} c_j} \leq \alpha_k$$

then $j^* = k$. Else, it is easy to check that

$$\frac{\sum_{j \leq k-1} c_j \alpha_j}{c + \sum_{j \leq k-1} c_j} > \alpha_k$$

2. As an inductive hypothesis, suppose that

$$\frac{\sum_{j \leq r} c_j \alpha_j}{c + \sum_{j \leq r} c_j} > \alpha_{r+1}.$$

If the LHS is less than or equal to α_r, set $j^* = r$, and STOP. Else, set $r = r - 1$ and repeat induction.

It is easy to see that this algorithm will stop as,

$$\frac{c_1\alpha_1}{c+c_1} < \alpha_1.$$

3.3 Conclusion and Ongoing Work

In this paper we modelled portfolio credit risk as an evolving function of time. The default probability of any obligor at any time depended on common systemic covariates, class dependent covariate and idiosyncratic random variables - we allowed a fairly general representation of conditional default probabilities that subsumes popular logit, default intensity based representations, as well as threshold based Gaussian and related copula models for defaults. The evolution of systemic covariates was modelled as a VAR(1) process. The evolution of class dependent covariates was modelled as an independent AR process (independent of systemic and other class covariates). We further assumed that these random variables had a Gaussian distribution. In this framework we analyzed occurrence of large losses as a function of time. In particular, we characterized the large deviations rate function of large losses. We also observed that this rate function is independent of the representation selected for conditional default probabilities.

This was a short note meant to highlight some of the essential issues. In our ongoing effort we build in more realistic and practically relevant features including:

1. We conduct large deviations analysis
    a. when the class and the systemic covariates are dependent with additional relaxations including allowing exposures and recoveries to be random. Further, as in Duffie et al. (2007), we also model firms exiting due to other reasons besides default, e.g., due to merger and acquisitions. We also allow defaults at any time to explicitly depend upon the level of defaults occurring in previous time periods (see, e.g., Sirignano and Giesecke 2015).
    b. when the covariates are allowed to have more general fatter-tailed distributions.
    c. when the portfolio composition is time varying.
2. Portfolio large loss probabilities tend to be small requiring massive computational effort in estimation when estimation is conducted using naive Monte Carlo. Fast simulation techniques are developed that exploit the large deviations structure of large losses (see, e.g., Juneja and Shahabuddin 2006; Asmussen and Glynn 2007 for introduction to rare event simulation).

Appendix: Some Proofs

Let $||x||^2 = \sum_{i=1}^{n} x_i^2$. Consider the optimization problem

$$\min ||x||^2 \tag{3.13}$$

$$s.t. \sum_{j=1}^{n} a_{i,j} x_j \geq b_i \quad i = 1, \ldots, m, \tag{3.14}$$

and let x^* denote the unique optimal solution of this optimization problem. It is easy to see from first order conditions that if $(b_i : i \leq m, b_i > 0)$, is replaced by $(\alpha b_i : i \leq m), \alpha > 0$, then the solution changes to αx^*.

Let $(X_i : i \leq n)$ denote i.i.d. Gaussian mean zero variance 1 random variables and let $d(n)$ denote any increasing function of n such that $d(n) \to \infty$ as $n \to \infty$.

The following lemma is well known and stated without proof (see, e.g., Glasserman et al. 2007).

Lemma 3.3 *The following holds:*

$$\lim_{n\to\infty} \frac{1}{d(n)} \log P\left(\sum_{j=1}^{n} a_{i,j} X_j \geq b_i d(n) + o(d(n)) \; i = 1, \ldots, m\right) = -||x^*||^2.$$

Proof of Lemma 3.1: Recall that we need to show that

$$\lim_{n\to\infty} \frac{1}{r_n^2} \log P(\mathscr{H}) = -q^*(t). \tag{3.15}$$

where $P(\mathscr{H})$ denotes the probability of the event that

$$\sum_{j=1}^{t}\sum_{k=1}^{d} h_{t-j,k} E_{j,k} + \sum_{k=1}^{t} \eta^{t-k} \Lambda_{1,k} \geq r_n \alpha_1 + r_n^\delta$$

and

$$\sum_{j=1}^{\tilde{t}}\sum_{k=1}^{d} h_{\tilde{t}-j,k} E_{j,k} + \sum_{k=1}^{\tilde{t}} \eta^{\tilde{t}-k} \Lambda_{1,k} \leq r_n \alpha_1 - r_n^\delta$$

for $1 \leq \tilde{t} \leq t - 1$.

From Lemma 3.3, to evaluate (3.15), it suffices to consider the optimization problem (call it **O1**),

$$\min \sum_{k=1}^{t}\sum_{p=1}^{d} e_{k,p}^2 + \sum_{k=1}^{t} l_k^2 \tag{3.16}$$

$$\text{s. t.} \sum_{k=1}^{t}\sum_{p=1}^{d} h_{t-k,p}e_{k,p} + \sum_{k=1}^{t}\eta_1^{t-k}l_k \geq \alpha_1, \tag{3.17}$$

$$\text{and} \sum_{k=1}^{\tilde{t}}\sum_{p=1}^{d} h_{\tilde{t}-k,p}e_{k,p} + \sum_{k=1}^{\tilde{t}}\eta_1^{\tilde{t}-k}l_k \leq \alpha_1. \tag{3.18}$$

for $1 \leq \tilde{t} \leq t-1$.

We first argue that in **O1**, under the optimal solution, the constraints (3.18) hold as strict inequalities.

This is easily seen through a contradiction. Suppose there exists an optimal solution $(\hat{e}_{k,p}, \hat{l}_k, k \leq t, p \leq d)$ such that for $\hat{t} < t$,

$$\sum_{k=1}^{\hat{t}}\sum_{p=1}^{d} h_{\hat{t}-k,p}\hat{e}_{k,p} + \sum_{k=1}^{\hat{t}}\eta_1^{\hat{t}-k}\hat{l}_k = \alpha_1$$

and if $\hat{t} > 1$, then for all $\tilde{t} < \hat{t}$ (3.18) are always strict. We can construct a new feasible solution with objective function at least as small with the property that constraints (3.18) are always strict.

This is done as follows: Let $s = t - \hat{t}$. Set $\bar{e}_{k+s,p} = \hat{e}_{k,p}$ for all $k \leq \hat{t}$ and $p \leq d$. Similarly, set $\bar{l}_{k+s} = \hat{l}_k$ for all $k \leq \hat{t}$. Set the remaining variables to zero.

Also, since the variables $(\bar{e}_{k,p}, \bar{l}_k k \leq t, p \leq d)$ satisfy constraint (3.18) with variables $(\bar{e}_{k,p}, \bar{l}_k k \leq s, p \leq d)$ set to zero, the objective function can be further improved by allowing these to be positive. This provides the desired contradiction. The specific form of $q^*(t)$ follows from the straightforward observation in Remark 3.4. □.

Proof of Theorem 3.1:

Now,

$$P(\mathcal{N}) \geq P(\mathcal{N}|\mathcal{H}_1^\tau)P(\mathcal{H}_1^\tau).$$

We argue that $P(\mathcal{N}|\mathcal{H}_1^\tau)$ converges to 1 as $n \to \infty$. This term equals

$$P\left(\frac{N_1(\tau)}{n} \geq a_\tau, \frac{\sum_{t=1}^{\tau-1} N_1(t)}{n} \leq c_1 - a_\tau|\mathcal{H}_1^\tau\right).$$

This may be further decomposed as

$$P\left(\frac{\sum_{t=1}^{\tau-1} N_1(t)}{n} \leq c_1 - a_\tau|(\cap_{t=1}^{\tau-1}\tilde{\mathcal{H}}_{1,t})\right) \tag{3.19}$$

times

$$P\left(\frac{N_1(\tau)}{n} \geq a_\tau|\frac{\sum_{t=1}^{\tau-1} N_1(t)}{n} \leq c_1 - a_\tau, \mathcal{H}_{1,\tau}\right). \tag{3.20}$$

To see that (3.19) converges to 1 as $n \to \infty$, note that it is lower bounded by

$$1 - \sum_{t=1}^{\tau-1} P\left(\frac{N_1(t)}{n} \geq \varepsilon | (\cap_{t=1}^{\tau-1} \tilde{\mathscr{H}}_{1,t})\right)$$

for $\varepsilon = (c_1 - a_\tau)/(\tau - 1)$. Consider now,

$$P\left(\frac{N_1(1)}{n} \geq \varepsilon | \tilde{\mathscr{H}}_{1,1}\right)$$

This is bounded from above by

$$2^{c_1 n} P(Z_{1,1}(n) \geq r_n^\delta)^{\varepsilon n}$$

where $2^{c_1 n}$ is a bound on number of ways at least εn obligors of Class 1 can be selected from $c_1 n$ obligors. Equation 3.19 now easily follows.

To see (3.19), observe that this is bounded from above by

$$2^{c_1 n} P(Z_{i,\tau}(n) \leq -r_n^\delta)^{(c_1 - a_\tau)n}$$

Since this decays to zero as $n \to \infty$, (3.19) follows.

In view of Lemma 3.1, we then have that

$$\lim_{n\to\infty} \frac{1}{r_n^2} \log P(\mathscr{N} \cap \mathscr{H}_1^\tau) = -q^*(\tau),$$

and thus large deviations lower bound follows. To achieve the upper bound, we need to show that

$$\limsup_{n\to\infty} \frac{1}{r_n^2} \log P(\mathscr{N}) \leq -q^*(\tau). \tag{3.21}$$

Observe that

$$P(\mathscr{N}) \leq P(H_\tau + Y_{1,\tau} \geq r_n\alpha_1 - r_n^\delta) + P\left(\frac{N_1(\tau)}{n} \geq a_\tau, H_\tau + Y_{1,\tau} \leq r_n\alpha_1 - r_n^\delta\right).$$

Now, from Lemma 3.3 and proof of Lemma 3.1,

$$\lim_{n\to\infty} \frac{1}{r_n^2} \log P(H_\tau + Y_{1,\tau} \geq r_n\alpha_1 - r_n^\delta) = -q^*(\tau).$$

Now,

$$P\left(\frac{N_1(\tau)}{n} \geq a_\tau, H_\tau + Y_{1,\tau} \leq r_n\alpha_1 - r_n^\delta\right)$$

is bounded from above by

$$2^n P(Z_{i,\tau} > r_n^\delta)^{na_\tau}$$

so that due to Assumption 2,

$$\limsup_{n\to\infty} \frac{1}{r_n^2} \log P\left(\frac{N_1(\tau)}{n} \geq a_\tau, H_\tau + Y_{1,\tau} \leq r_n\alpha_1 - r_n^\delta\right) = -\infty,$$

and (3.21) follows. □

References

Asmussen, S. and Glynn, P.W., 2007. *Stochastic simulation: algorithms and analysis* (Vol. 57). Springer Science & Business Media.

Bassamboo, A., Juneja, S. and Zeevi, A., 2008. Portfolio credit risk with extremal dependence: Asymptotic analysis and efficient simulation. *Operations Research*, **56** (3), pp. 593–606.

Chava, S. and Jarrow, R.A., 2004. Bankruptcy prediction with industry effects. *Review of Finance*, **8**(4), pp. 537–569.

Dembo, A., Deuschel, J.D. and Duffie, D., 2004. Large portfolio losses. *Finance and Stochastics*, **8** (1), pp. 3–16.

Dembo, A. and Zeitouni, O., 2009. *Large deviations techniques and applications*. Vol. 38. Springer Science & Business Media, 2009.

Duan, J.C. and Fulop, A., 2013. Multiperiod Corporate Default Prediction with the Partially-Conditioned Forward Intensity. Available at SSRN 2151174.

Duan, J.C., Sun, J. and Wang, T., 2012. Multiperiod corporate default prediction?A forward intensity approach. *Journal of Econometrics*, **170** (1), pp. 191–209.

Duffie, D., Saita, L. and Wang, K., 2007. Multi-period corporate default prediction with stochastic covariates. *Journal of Financial Economics*, **83** (3), pp. 635–665.

Duffie, D. and Singleton, K.J., 2012. *Credit risk: pricing, measurement, and management*. Princeton University Press.

Giesecke, K., Longstaff, F.A., Schaefer, S. and Strebulaev, I., 2011. Corporate bond default risk: A 150-year perspective. *Journal of Financial Economics*, **102** (2), pp. 233–250.

Glasserman, P. and Li, J., 2005. Importance sampling for portfolio credit risk. *Management Science*, **51** (11), pp. 1643–1656.

Glasserman, P., Kang, W. and Shahabuddin, P., 2007. Large deviations in multifactor portfolio credit risk. *Mathematical Finance*, **17** (3), pp. 345–379.

Juneja, S., and Shahabuddin, P., 2006. Rare-event simulation techniques: an introduction and recent advances. *Handbooks in operations research and management science*, **13**, 291–350.

Merton, R.C., 1974. On the pricing of corporate debt: The risk structure of interest rates. *The Journal of Finance*, **29** (2), pp. 449–470.

Sirignano, J. and Giesecke, K. 2015. Risk analysis for large pools of loans. Available at SSRN 2514040, 2015

Zhang, X., Blanchet, J., Giesecke, K. and Glynn, P.W., 2015. Affine point processes: approximation and efficient simulation. *Mathematics of Operations Research*, **40** (4), pp. 797–819.

Chapter 4
Extreme Eigenvector Analysis of Global Financial Correlation Matrices

Pradeep Bhadola and Nivedita Deo

Abstract The correlation between the 31 global financial indices from American, European and Asia-Pacific region are studied for a period before, during and after the 2008 crash. A spectral study of the moving window correlations gives significant information about the interactions between different financial indices. Eigenvalue spectra for each window is compared with the random matrix results on Wishart matrices. The upper side of the spectra outside the random matrix bound consists of the same number of eigenvalues for all windows where as significant differences can be seen in the lower side of the spectra. Analysis of the eigenvectors indicates that the second largest eigenvector clearly gives the sectors indicating the geographical location of each country i.e. the countries with geographical proximity giving similar contributions to the second largest eigenvector. The eigenvalues on the lower side of spectra outside the random matrix bounds changes before during and after the crisis. A quantitative way of specifying information based on the eigenvectors is constructed defined as the "eigenvector entropy" which gives the localization of eigenvectors. Most of the dynamics is captured by the low eigenvectors. The lowest eigenvector shows how the financial ties changes before, during and after the 2008 crisis.

4.1 Introduction

The economic and social growth of a country depends on the state of its financial market (Lin et al. 2012). Financial markets are very complex to understand, having many unidentified factors and interactions that govern their dynamics. Even with the enormous growth of the financial data and with the increase in the computational capabilities by developing the high-throughput methods, to understand the complex behavior of the financial market remains a great challenge. The studies on the financial

P. Bhadola · N. Deo (✉)
Department of Physics and Astrophysics, University of Delhi, Delhi 110007, India
e-mail: ndeo@physics.du.ac.in

P. Bhadola
e-mail: bhadola.pradeep@gmail.com

F. Abergel et al. (eds.), *Econophysics and Sociophysics: Recent Progress and Future Directions*, New Economic Windows,
DOI 10.1007/978-3-319-47705-3_4

markets either by analyzing enormous financial data or by modeling the behavior of the financial market implies that the market shows non-equilibrium properties. The use of the cross correlations between the financial data have been extensively studied at different time scales (Conlon et al. 2007, 2008; Podobnik and Stanley 2008; Kenett et al. 2015). Random matrix theory (RMT) have been extensively used for filtering of the spectrum of financial correlation matrix to separate the relevant information from the noise. The eigenvalues of the correlation matrix which deviated significantly from the RMT predictions provides crucial information about the interaction and structure of the financial markets. Studies are mostly concerned about the eigenvalues which deviates from RMT upper bound of the eigenvalue spectrum. But many systems such as biological have shown that the low eigenvalues contains useful information (Cocco et al. 2013) and are very useful in predicting the clusters or sectors (Pradeep Bhadola and Deo 2016).

This work aims at the use of the eigenvector based method to infer the vital information about a system. We have used a method called eigenvector localization to extract the community structure and interaction among different agents in a system. Eigenvector localization refers to the condition where most of the weight is association with only few eigenvector components. For instance, in case of a large graph, the eigenvector components of some of the extreme eigenvectors of the Adjacency matrix (eigenvector corresponding to extreme eigenvalues) will have most of the weight concentrated on the nodes with very high degree. In this paper we are using the eigenvector localization of the correlation of world financial indices to derive meaningful clusters of financial interaction among countries.

4.2 System and Data

For the analysis, we use the daily adjusted closing stock price of the 31 financial market representing different region of the world from beginning of 2006 to the end of 2015. The 31 indices follows the European market, Asian market and American etc. On a particular day if 50 % of the indices observed a holiday then we removed that day from our analysis thus only considering days when 50 % or more indices are active.

If $S_i(t)$ is the price of country index i at time t, then the logarithmic returns $R_i(t)$ is calculated as

$$R_i(t) = ln(S_i(t + \Delta t)) - ln(S_i(t)) \tag{4.1}$$

where the time lag $\Delta t = 1$ day. The normalized returns is given by

$$r_i(t) = \frac{R_i(t) - < R_i >}{\sigma_i} \tag{4.2}$$

where $< R_i >$ is the time average of the returns over the time period and σ_i is the standard deviation of $R_i(t)$ defined as $\sigma_i = < R_i^2 > - < R_i >^2$.

Pearson's correlation coefficient is used to estimate the correlations present among different financial indices. The correlation matrix between the stocks is defined as

$$C_{i,j} = < r_i(t) r_j(t) > . \tag{4.3}$$

The correlation obtained are such that $-1 \leq C_{i,j} \leq 1$ where $C_{i,j} = 1$ represents perfect correlation and $C_{i,j} = -1$ represents perfect anti-correlation.

The eigenvalue equation $C\hat{v_i} = \lambda_i \hat{v_i}$ is used to determine the eigenvalues λ_i and eigenvectors v_i of the correlation matrix C. The eigenvalues are arranged in the ascending order of magnitude such that $\lambda_1 \leq \lambda_2 \leq \lambda_3 \cdots \leq \lambda_N$.

We have studied the financial data with a moving window correlation. The size of the window is 250 days with a shift of 100 days. The correlation matrix is calculated for each window and the properties are studied.

The null model is constructed by randomly shuffling the data for each index so to remove any existing correlations. Numerically, the correlation for the shuffled system are equivalent to the Wishart matrices. Therefore, the analytical results of the Wishart matrices are used to compare the spectral properties of the financial correlation matrix. The spectral properties of Wishart matrices are well defined (Bowick and Brezin 1991) where the eigenvalue density function $P_W(\lambda)$ is given by

$$P_W(\lambda) = \frac{Q}{2\pi\sigma^2} \frac{\sqrt{(\lambda_+ - \lambda)(\lambda - \lambda_-)}}{\lambda} . \tag{4.4}$$

with $Q = \frac{W}{L} \geq 1$ and $\sigma = 1$ the standard deviation.

The distribution is known as Marcenko-Pastur distribution where the upper and lower bounds for the eigenvalue λ are given by

$$\lambda_{\pm} = \sigma^2 \left(1 + \frac{1}{Q} \pm 2\sqrt{\frac{1}{Q}} \right) \tag{4.5}$$

For the current analysis $W = 250$ is the size of the window and $N = 31$ is the number of indices used for the study which gives $Q = 8.064$. The lower and upper bounds on the eigenvalues imposed by the random matrix theory is thus $\lambda_+ = 1.82$ and $\lambda_- = 0.419$.

Analyzing the eigenvalue distribution for various windows Figs. 4.1 and 4.2 shows the distribution of eigenvalues larger than the random matrix bounds for ($\lambda \geq \lambda_+$) are nearly identical for all windows. But the distribution of eigenvalues lower than lower RMT bound ($\lambda \leq \lambda_-$) is different for all windows. The results are shown for window 2 (25-May-2006 to 15-May-2007, calm period), window 3 (12-Oct-2006 to 02-Oct-2007, Onset of crash), window 7 (06-May-2008 to 23-April-2009, during crash), window 18 (17-Aug-2012 to 26-Jul-2013, after crash) and window 24 (12-May-2014 to 25-Nov-2015, after crash).

On an average there are only two eigenvalues outside the RMT bound on the upper side ($\lambda \geq \lambda_+$) but there are more than 15 eigenvalues lower than the lower

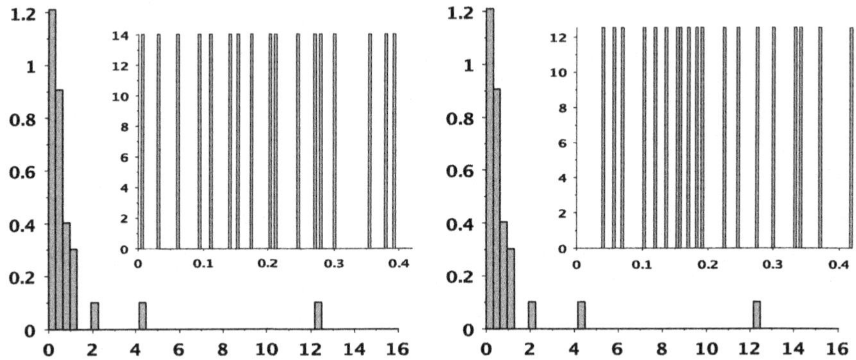

Fig. 4.1 Eigenvalue distributions. (*Left*) EV distribution for window 2 (calm period). (*Right*) EV distribution for window 3 (onset of crash). Insets show EVs outside the lower bound

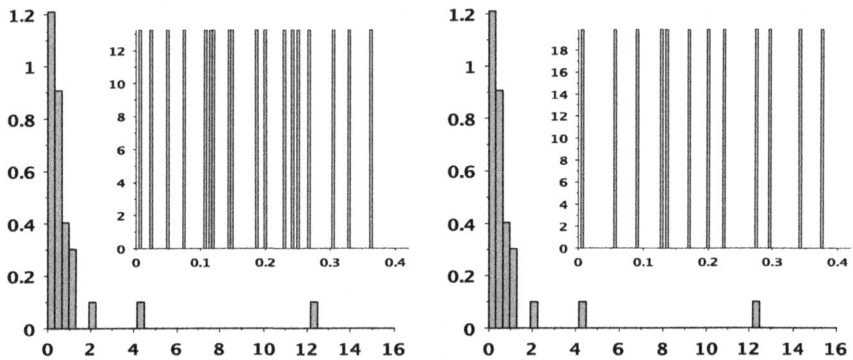

Fig. 4.2 Eigenvalue distributions. (*Left*) EV distribution for window 7 (During Crash). (*Right*) EV distribution for window 18 (after Crash). Insets show EVs outside the lower bound

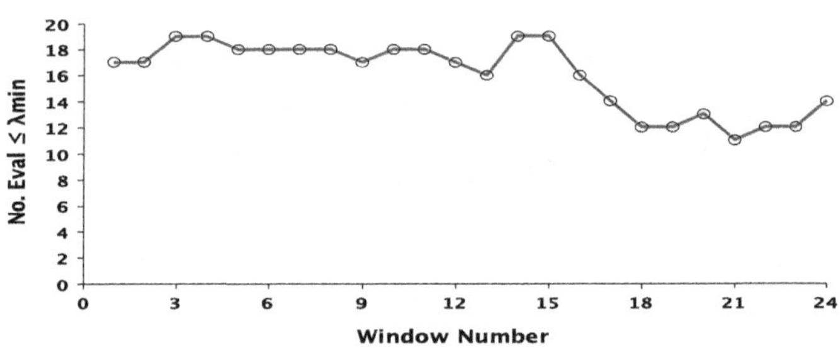

Fig. 4.3 Outside RMT

RMT bound ($\lambda \leq \lambda_-$) see Fig. 4.3. Hence most of the information which is outside the RMT bound is located on the lower side of the spectrum. Thus the essence of the interaction among agents is indicated mostly by the eigenvalues which are outside the lower RMT bound.

4.3 Eigenvalue Dynamics

The time evolution of the eigenvalues of the correlation matrix is studied. The Fig. 4.4 shows the time evolution of eigenvalues of the correlation matrix for a sliding window of 250 days. The dynamics of the first few smallest eigenvalues are opposite to the largest and second largest eigenvalues resulting in eigenvalue repulsion. The eigenvalue repulsion between the largest and the sum of the smallest 25 eigenvalues is shown in Fig. 4.5 where the sum of the few eigenvalues is opposite in direction with time to those of the largest eigenvalues. The eigenvalues which are inside the RMT bound have very small fluctuations with time. Thus the dynamics indicates that the information of change is mostly contained in the eigenvalues outside the RMT bound.

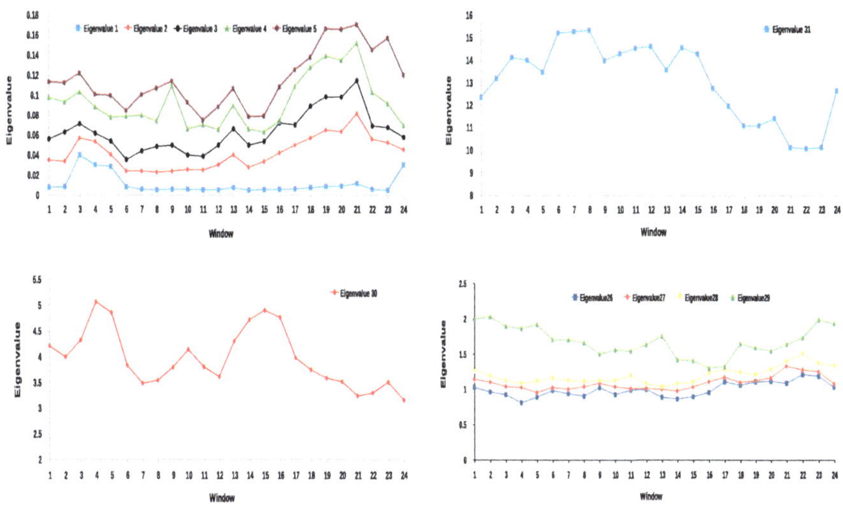

Fig. 4.4 Time evolution of the first four smallest eigenvalues (outside lower bound), the largest, second largest and some eigenvalues inside the RMT bound are shown

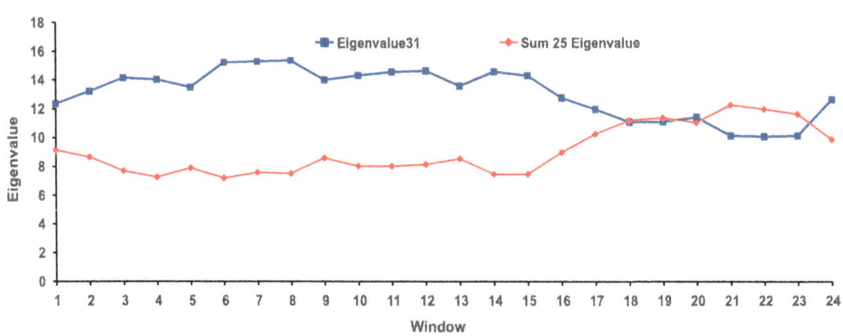

Fig. 4.5 Time evolution of the sum of the smallest 25 eigenvalues and the largest eigenvalues

4.4 Localization of Eigenvectors

Shannon's entropy is used to estimate the information content of each eigenvector. This estimation gives the information about the localization of the eigenvector components. The entropy of an eigenvector is defined as

$$H_i = -\sum_{j=1}^{L} u_i(j) log_L(u_i(j)), \tag{4.6}$$

where L is the total number of indices (number of eigenvector components) and $u_i(j) = (v_i(j))^2$ is the square of the jth component of the ith normalized vector v_i.

The eigenvector entropy shows that the low eigenvectors are highly localized and are suitable candidates for the estimation of the strong interactions among different agents. As, for some biological systems, especially in the spectral analysis of correlations between positions for a protein family, it is found that low eigenmodes are more informative (Cocoo et al. 2013) and gives useful insight about the interactions, sectors and the community structure present within the system (Pradeep Bhadola and Deo 2016). In the literature of financial analysis it is shown, in Markowitzs theory of optimal portfolios (Elton et al. 1995/1959/1997), that the lowest eigenvalues and eigenvectors are significant and important for the system which physically represents the least risky portfolios (Elton and Gruber 1995/1959/1997).

Figure 4.6, clearly shows that difference between the eigenvector entropy between different parts of the spectra. The entropy of the small eigenvectors are low as compared to large eigenvector. The highly localized eigenvectors gives a sector or cluster by collecting the components with significant contribution in the entropy versus the eigenvector plot. To extract the indices contributing to the cluster we use the square of the eigenvector, which is the net contribution from that component towards that sector. Components with a high contribution forms a group of very close financial ties.

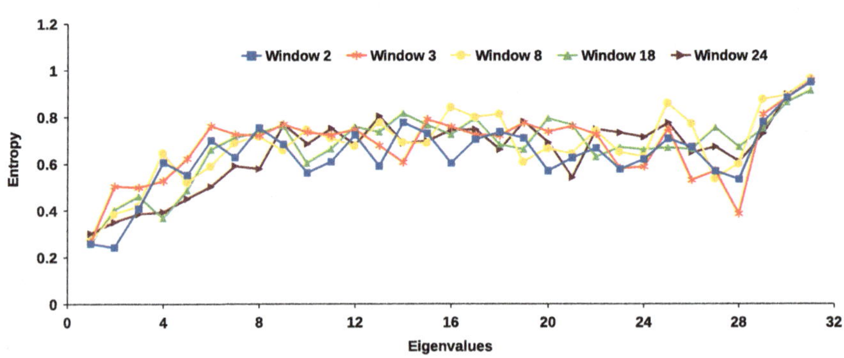

Fig. 4.6 Entropy of each eigenvectors for different windows

4.5 Analysis of Eigenvectors

4.5.1 *Largest Eigenvector*

The eigenvalues on both side of the spectrum shows significant differences from the RMT predictions. The eigenvectors corresponding to these eigenvalues should contain non random information present in the system. Analysis of the eigenvectors corresponding to the largest eigenvalues over the time window shows that the European countries are financially the most active countries by contributing most to the largest eigenvector. Since largest eigenvector shows the state of the system as a whole and in the present case represents the global economic conditions, therefore analysis of components indicates that the European market mainly decides the global economic conditions. Some of the Asian countries have small contribution towards the global economic state. These results are true for all the windows analyzed (Fig. 4.7).

4.5.2 *Second Largest Eigenvector*

The components of the second largest eigenvectors is shown in Fig. 4.8 for different windows. The analysis of components indicates that the global market can be divided broadly into two categories depending on the sign of component. Categorizing the components into these two categories we find that one group is the European and American market where as the other group with opposite sign comprises the countries from Asia-Pacific region. These members of the group do not change with time and is constant across all windows only a flip in sign is observed for some windows. Thus the second largest eigenvector physically categorizes countries based on their

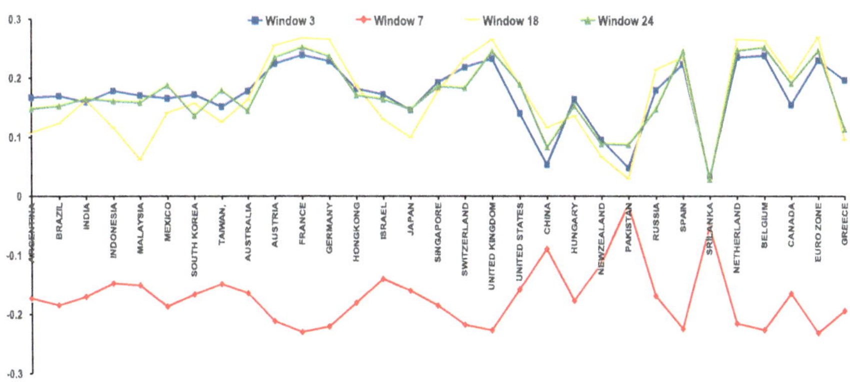

Fig. 4.7 Components of the largest eigenvectors for different windows

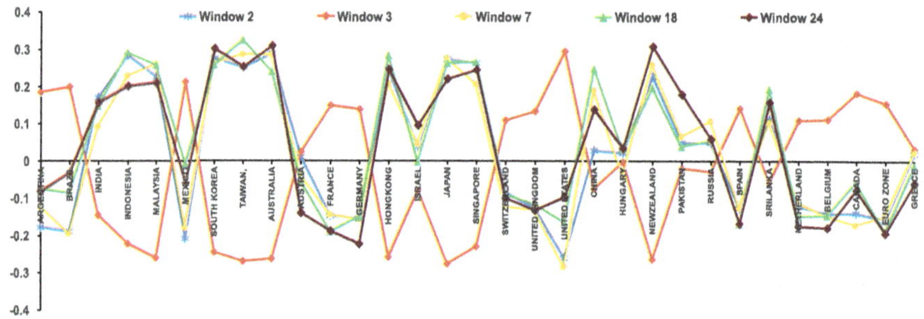

Fig. 4.8 Components of the second largest eigenvectors for different windows

Table 4.1 Sector based on the second largest eigenvector specifying geographical location of countries

Sector	Geographic location	Countries
1	European and American	Argentina, Brazil, Mexico, Austria, France, Germany, Switzerland, UK, US, Hungary, Spain, Netherlands, Belgium, Canada, Greece and Euro Zone (EURO STOXX 50 index)
2	Asia Pacific	India, Indonesia, Malaysia, South Korea, Taiwan, Australia, Hong Kong, Israel, Japan, Singapore, China, New Zealand, Pakistan, Russia, Sri Lanka

geographical locations (showing that geographical proximity corresponds to strong interactions). Table 4.1 gives the sectors and the geographical location of countries as obtained by analyzing the second largest eigenvector.

4.5.3 Smallest Eigenvector

Eigenvectors on the lower side of the spectra are highly localized and contain useful information. Analyzing the components of the smallest eigenvector Fig. 4.9 shows a change in weights of components during the crash. France, Germany, Spain, Belgium and Euro zone which were dominant during the calm period, window 2 (25-May-2006 to 15-May-2007) but just before the crash, that is, for window 3 (12-Oct-2006 to 02-Oct-2007) there is a major shift in contribution and France, UK, Netherland and Euro zone (EURO STOXX 50 index) now have the dominant contribution. Germany, Spain and Belgium have almost zero contribution just before the crash, see window 3 (12-Oct-2006 to 02-Oct-2007), where as France and Euro zone (EURO STOXX 50 index) have a reduced effective contribution. UK and Netherland have higher contribution just before the crash, see window 3 (12-Oct-2006 to 02-Oct-2007).

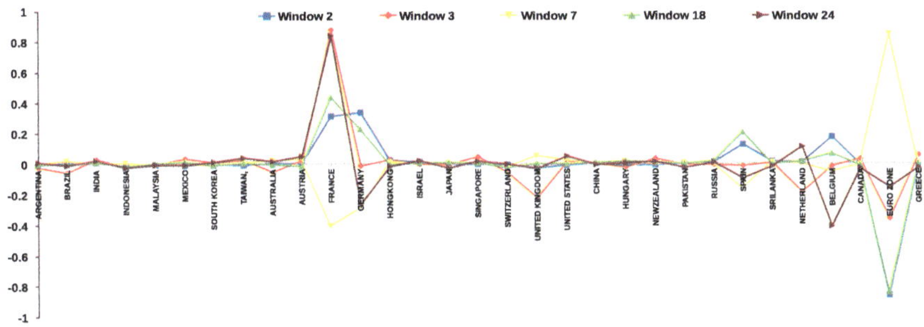

Fig. 4.9 Components of the smallest eigenvectors for different windows

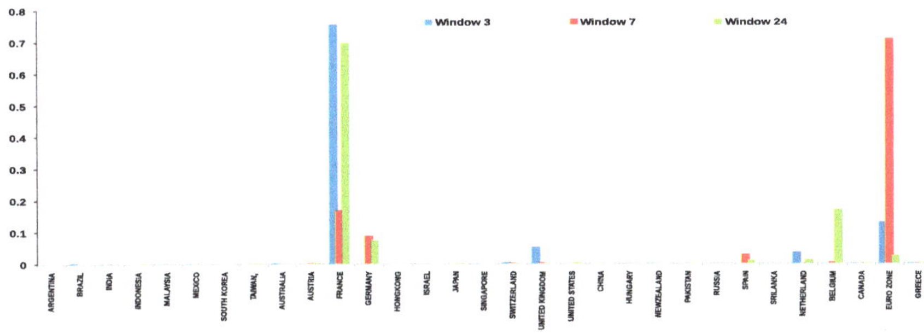

Fig. 4.10 Square of the components of the smallest eigenvectors for different windows

During the crash, see window 7 (06-May-2008 to 23-April-2009), France, Germany, UK, Spain, Belgium and Euro zone (EURO STOXX 50 index) have significant contributions to the smallest eigenvector.

The sector during the calm period (window 2) consists of France, Germany, Spain, Belgium and Euro zone which changes to France, UK, Netherland and Euro zone just before the crisis (window 3). During the crisis there is again a shift in countries (change in the sector) comprising of France, Germany, Spain, Belgium and Euro zone. This shift in the financial ties maybe responsible for the global crisis. After the crash France, Germany, Spain, Belgium and Euro zone remain in the sector for the rest of the period.

The effective contribution from each country can be clearly seen in Fig. 4.10 where the square of the component is plotted for different windows. The change in contribution of components towards the smallest eigenvector can be clearly seen before crisis (window 3), during crisis (window 7) and after the crisis (window 24). This change in contribution from the countries may be linked to the onset of crisis as there is significant change in financial ties between the countries.

This establishes that most of the dynamics is captured by the eigenvectors corresponding to the smallest eigenvalues for financial global indices before, during and after the 2008 crisis.

4.6 Conclusion

In this manuscript, the global financial market is studied from 2006 to 2015 by construction a moving window correlations with a window size of 250 days and shift of 100 days. The 2008 crash is studied in details and the change in the correlation between different countries over time is studied. The eigenvalue distribution for all windows is created and compared with the random matrix results. The eigenvalues of the financial matrix greater than the upper limit of the RMT bound shows identical behavior for all windows. The distribution of eigenvalues on the lower side of spectra outside the RMT lower bound shows significant changes over different windows. Most of the information is located in the lower side of the spectra. The second largest eigenvalues is linked to the geographical linking of countries and the structure of the second largest eigenvector components remains the same for all windows only a flip in sign of the component is observed. These eigenvalues can be further used to unveil significant Information. A measure known as eigenvector entropy is introduced to check the localization of each eigenvalue. Eigenvalues on the lower side of the spectra are more localized as compared to the eigenvalues on the upper side. Analyzing the smallest eigenvalues indicates a change in financial ties before, during and after the crash and hence may throw light on the reason of the crash.

Acknowledgements We acknowledge Delhi Universit R&D Grant for financial support.

References

Pradeep Bhadola & N. Deo, Targeting functional motif in a protein family , Submitted to journal.

Bowick M. J. and Brezin E., Phys. Lett. B **268**, 21 (1991); Feinberg J. and Zee A., J. Stat. Phys. **87**, 473 (1997).

Simona Cocco, Remi Monasson and Martin Weigt. PLoS Computational Biology **9** (**8**), e1003176 (2013).

Cocco, S., Monasson, R. & Weigt, M. From principal component to direct coupling analysis of coevolution in proteins: low-eigenvalue modes are needed for structure prediction. PLoS Comput. Biol. 9, e1003176 (2013).

Conlon, T., Ruskin, H.J., Crane, M., Wavelet multiscale analysis for Hedge Funds: Scaling and strategies, To appear: Physica A (2008), doi:10.1016/j.physa.2008.05.046

Conlon, T., Ruskin, H.J., Crane, M., Random matrix theory and fund of funds portfolio optimisation, Physica A 382 (2) (2007) 565-576.

E.J. Elton and M.J. Gruber, Modern Portfolio Theory and Investment Analysis (J.Wiley and Sons, New York, 1995); H. Markowitz, Portfolio Selection: Efficient Diversification of Investments (J.Wiley and Sons, New York, 1959). See also: J.P. Bouchaud and M. Potters, Theory of Financial Risk, (Alea-Saclay, Eyrolles, Paris, 1997) (in French).

Kenett, D. Y., Huang, X., Vodenska, I., Havlin, S., & Stanley, H. E. (2015). Partial correlation analysis: Applications for financial markets. Quantitative Finance, 15(4), 569-578.
Lin, C. S., Chiu, S. H., & Lin, T. Y. Empirical mode decompositionbased least squares support vector regression for foreign exchange rate forecasting. Economic Modelling, 29(6), 2583-2590 (2012).
Podobnik, B., Stanley, H.E., Detrended cross-correlation analysis: A new method for analysing two non-stationary time series, Phys. Rev. Lett 100 (2008) 084102.

Chapter 5
Network Theory in Macroeconomics and Finance

Anindya S. Chakrabarti

Abstract In the last couple of years, a large body of theoretical and empirical work has been done emphasizing the network structures of economies. Availability of large data and suitable tools have made such a transition possible. The implications of such work is also apparent in settling age-old debates of micro versus macro foundations and to model and quantify shock propagation mechanisms through the networks. Here I summarize a number of major topics where significant work has been done in the recent times using both physics-based and economics-based models of networks and show that they are complementary in approach.

5.1 Introduction

Economic systems are complex (Krugman 1996). With increasing amount of interdependence among a vast number of entities, earlier paradigms in the prevailing economic theory often fall short of providing a close approximation of the reality. Schweitzer et al. (2009) argues that there is a 'critical need' for a theoretical tool to understand and model that complexity and may be *network theory* provides that tool. The goal is to understand propagation, diversification and aggregation of shocks and spill-over effects of economic quantities, which will potentially help us to tackle a wide range of problems: from assessment of risk embedded in a financial network to understanding contractions and expansions of global trade flow when a recession hits one country. In the same vein, Buchanan (2013) argues that the complex network approach might prove very useful for understanding the financial world.

In this short note, I summarize some basic tools and models in graph (network) theory and describe the findings of an emerging stream of literature in macroeconomics and finance, that studies economic and financial networks. Let us begin by defining a number of important concepts and variables in graph theory. Then we can go on to discuss several features real world graphs and some theoretical underpinnings.

A.S. Chakrabarti (✉)
Economics Area, Indian Institute of Management, Vastrapur 380015, Ahmedabad, India
e-mail: anindyac@iimahd.ernet.in

F. Abergel et al. (eds.), *Econophysics and Sociophysics: Recent Progress and Future Directions*, New Economic Windows,
DOI 10.1007/978-3-319-47705-3_5

There are many expositions on graph theory and their relations with real world data (e.g. Barrat et al. 2008; Jackson 2010; Caldarelli 2007). However, macroeconomic networks are rarely discussed in those textbooks.

We define a graph as collection of nodes and edges.

Definition 1 A graph $G = (N, E)$ is defined as a set of finite number of nodes N and edges E.

Intuitively, the number of nodes represents the size of a graph. However, traditionally the graph theorists have defined size by cardinality of the set E (see Barrat et al. 2008). Usually researchers do not differentiate between the terms *graph* and *network*. But at the same time, to describe properties of a comparatively smaller set of nodes, people use the former whereas for describing connections between a large number of nodes, the latter term is used. Thus in pure mathematics, the most common term for describing connections is 'graph' (where the focus is on very specific properties of very specific edges and nodes; the aggregate is thought of as a multiplication of such small units) whereas in the applied sciences like physics or economics, usually such structures are referred to by the term 'network' (where the aggregate properties of the whole ensemble matter; node-specific properties are less important). Thus graphs are of essentially abstract mathematical nature whereas networks have a physical significance representing some type of physical connection between nodes through trade, information, migration etc. An example of the type of economic problems that we will discuss below, is as follows. Given that the distribution of degree (a quantity we will define below) of the input-output network has a fat-tail (fatter than say, log-normal), how much aggregate volatility can be explained by idiosyncratic shocks to the nodes in the tail?

Definition 2 The term 'network' $\mathbb{N}$ will be used to denote a graph with a statistically large number of nodes, N.

In some cases, the edges might not be symmetric between two nodes. For example, consider a citation network. Researchers are considered nodes and if the i-th researcher cites the j-th, we add an edge from i to j. However that edge is directed as the edge from j to i may not materialize.

Definition 3 A graph with directed edges between nodes is called a directed graph.

For the above mentioned citation network, the edges have binary weight, 0 or 1. But in general we can consider different weights for different edges, e.g. trade network.

Definition 4 A graph with weighted edges is called a weighted graph.

In general, most economic networks are directed and weighted (for example, trade network). Many social networks could be directed but not weighted (e.g. citation), weighted but undirected (e.g. correlation graph in stock markets; to be discussed below) or undirected and unweighted (e.g. Facebook friendship).

Given a network, it is not necessary that all nodes are neighbors to each other. In other words, we may not have a *fully connected* network. For any two nodes (i and j), there might exist a sequence of nodes which if followed will lead from node i to j.

Definition 5 A path is a sequence of nodes connecting two separate nodes.

Average path length (as the name suggests, it is the average of all path lengths) is an important characteristic of networks. For economics, probably the most important characteristic is connectivity or degree distribution. The degree of a node captures how many neighbors does one node have.

Definition 6 Degree or connectivity of a node of a network $\mathbb{N}$ is the number of edges incident to that node.

To see why it is important, consider a banking network. If the i-th bank lends to d_i number of other banks, we say that the degree of that node is d_i. Bigger this number is, usually the more important is this bank to its downstream partners for lending and also for the i-th bank itself, it also captures how exposed it is to failure because of the loans made.

However there are different ways to characterize importance of different nodes in a given network and degree centrality is one of them. There are The idea of *closeness centrality* depends on how far is one node from the rest of the nodes. A natural measure of it would be the sum of reciprocal of all distances.

Definition 7 Closeness centrality is defined as the sum of reciprocal of all shortest paths from node i to all other nodes $j \in N$ with a standard convention that $1/\infty = 0$.

Another related measure is the *betweenness centrality* which gives a quantitative measure of how many times a particular node falls on the paths between all other pairs of nodes.

Definition 8 Betweenness centrality of a node i is defined as the sum over all possible pairs (i, j) of ratios of total number of shortest paths between (i, j) passing through the node i and the total number of shortest paths between the same pair of nodes (i, j).

However, neither of the above two definitions are very useful for economic quantities although they are very important for social and web networks.

The most used centrality measure in economics goes by the name of 'eigenvector centrality'.

Definition 9 Eigenvector centrality of a set of nodes in a network corresponds to the dominant eigenvector of its adjacency matrix.

Related measures are Bonacich centrality, Katz centrality etc. Very interestingly, this particular centrality measure is intuitively related to many economic entities. For example, one can construct a large scale production network by using input-output data of a country. Then centrality of different sectors can be constructed using the inflow-outflow matrix. In fact this approach has been found to be extremely useful for multiple formal economic models. See Acemoglu et al. (2012) for an application.

5.2 Descriptive Models

There are many excellent books (Barrat et al. 2008; Jackson 2010) and monographs (Dorogovtsev and Mendes 2003) on network structures and properties from mathematical as well as statistical point of view. Below I describe three types of networks with different properties. The list is not exhaustive and far from complete. I chose them because these three presents three important benchmarks for describing a network and the last one has found a lot of applications in economic and financial networks.

5.2.1 *Random Networks*

This type of networks have been studied in great details. These have a simple but mathematically elegant structure (Erdos and Renyi 1959). Not many economic networks belong to this class though. A generative model is as follows:

1. Start with N nodes and no edges.
2. Fix a probability p.
3. There are ${}^{N}C_2$ possible edges.
4. Choose the i-th possible edge and it materializes with probability p. Do it for all possible edges.

This will generate a random graph with approximately pE^{max} edges for a large N, where E^{max} is the maximum number of edges possible.

5.2.2 *Small World Networks*

These are very important for studying social networks and was made famous by the story of 'six-degrees of separation'. The basic idea is that in many networks, a large number of nodes are not neighbors of each other (implying that a lot of edges are missing) but the average distance between any two nodes is quite small. In particular, the average path-length is proportional to the log of size of the network,

$$l \propto \log N. \tag{5.1}$$

A very well known algorithm for generating a network with small-world property was proposed by Watts and Strogatz (1998) which can be summarized as follows.

1. Fix a number of nodes N and an integer number of average degree $\bar{d}$ such that $N \gg \bar{d} \gg \log(N) \gg 1$.
2. Choose a probability parameter p.

3. Construct a regular circular network with those N nodes each having connection with d nodes, with $d/2$ connections on each side.
4. Pick one node n_i. Consider all edges with nodes j such that $i < j$. Rewire those edges, each with probability p. For rewiring choose any other (other than itself and nodes with already existing connections with n_i) node with equal probability.

This mechanism generates a network with small-world property. By altering the parameter p, one can generate a regular network ($p = 0$) in one extreme and a random network ($p = 1$) on the other. Below we discuss scale-free networks and a generative mechanism.

5.2.3 *Networks with Fat Tails*

This class of networks have proved useful in describing economic phenomena. Notably, the input-output network seems to have a fat tail (Acemoglu et al. 2012). There are multiple mechanisms for generating scale-free networks. A very simple model in discrete time was proposed by Albert and Barabasi (2002) which uses a preferential attachment scheme to generate a degree distribution which is scale-free. The basic algorithm is as follows.

1. Start with some nodes which are connected to each other. Say, we have N^{ini} number of nodes.
2. At every point of time add $n \leq N^{ini}$ node.
3. The new node will form a connection with the existing i-th node with probability $p_i^{new} = d_i / \sum_j d_j$.
4. This process continues ad infinitum.

This process generates a network with degree distribution

$$P(d_i) \sim d_i^{-3}. \tag{5.2}$$

It has several other interesting properties like clustering coefficient that can be pinned down mathematically and many variants of it have been proposed in the literature (see Barrat et al. 2008 for details). The most relevant property for our purpose that can be utilized in the context of economics is the degree distribution.

5.3 Large-Scale Economic Networks

Bak et al. (1993) was an early attempt to understand aggregate economic fluctuations in terms of idiosyncratic shocks to individual agents. This also asks the question that should we consider idiosyncratic shocks as a source of volatility? There were several irrelevance theorems proved which essentially had shown that idiosyncratic shocks tend to cancel each other in a multi-sector set-up, but the debate was not settled (see e.g. Dupor 1999; Horvath 1998). Newer results, both theoretical and empirical, came up later which showed that the answer could well be positive which we describe below.

5.3.1 Input-Output Structures

One of the first network structures studied in great details is the input-output network (Leontief 1936, 1947, 1986). The essential postulate is that an economy is made up of a number of sectors that are distinct in their inputs and outputs. Each sector buys inputs from every other sector (some may buy zero inputs from some particular sectors) and sells output to other sectors. The workers provide labor, earn wage and consume the output net of input supplies.

Thus if one sector receives a negative shock, it can potentially transfer the effects to all downstream sectors. The question is whether that will so dispersed that none of it would be seen in the aggregate fluctuations or not. Acemoglu et al. (2012) studied it directly in the context of an input-output structure and showed that the degree distribution is sufficiently skewed (fat-tailed; there are some sectors disproportionately more important than the rest) so that idiosyncratic shocks to those sectors do not die completely. This provides a theoretical solution to the debate (it also shows that this channel is empirically relevant).

A simple exposition of the model is as follows (for details see Acemoglu et al. 2012). There is an unit mass of households with utility function defined over a consumption bundle $\{c_i\}_{i \in N}$ as

$$u = \xi . \prod_{i \in N} (c_i)^{1/N} \tag{5.3}$$

where ξ is a parameter. The production function for each sector is such that it uses some inputs from other sectors (or at least zero),

$$x_i = (z_i l_i)^{\alpha} (\prod_{j \in N} x_{ij}^{\omega_{ij}})^{1-\alpha} \tag{5.4}$$

where z_i is an idiosyncratic shock to the i-th sector and $\{\omega_i\}_{j \in N}$ captures the indegrees of the production network. To understand how it captures the network structure of production, take logs on both sides to get

$$\log(x_i) = \alpha \log(z_i) + \alpha \log(l_i) + (1 - \alpha) \sum_{j \in N} \omega_{ij} \log x_{ij}. \tag{5.5}$$

Since all sectors are profit maximizing, their optimal choice of inputs (how much would they buy from other sectors) will depend in turn on the prices and how much they themselves are producing. The markets clear at the aggregate level. After substitution, we can rewrite the equation above as (x^* being the solution to the above equation)

$$\log(x^*) = F(\log(z), \log(x^*)) \tag{5.6}$$

which means that the aggregate output of one sector is a function of all productivity shocks and optimal outputs of all sectors. Hence, this becomes a recursive system. Acemoglu et al. (2012) shows that the final GDP can be expressed as

$$\text{GDP} = w \log(z) \tag{5.7}$$

where w is a weight vector. Thus the output of all sectors are functions of the vector of all idiosyncratic shocks.

Foerster et al. (2011) considered the same question and provided a (neoclassical multi-sector) model to interpret data. They showed that idiosyncratic shocks explain about half of the variation in industrial production during the great moderation (which basically refers to two-decades long calm period before the recession in 2007).

5.3.2 *Trade Networks*

A very simple trade model can be provided on the basis of the input-output model presented above. Suppose there are N countries and for the households of the i-th country, the utility function is given by the same simple form:

$$u_i = \xi_i . \prod (c_i)^{1/N} \tag{5.8}$$

Each country gets an endowment of country-specific goods (e.g. Italian wine, German cars):

$$y_i = z_i \tag{5.9}$$

where z_i is a positive random variable. Now we can assume that there is a Walrasian market (perfectly competitive) across all countries. Each country is small enough so that it takes the world price as given. The we can solve the trading problem and can generate a trade flow matrix where weight of each edge is a function of the preference and productivity parameters,

$$e_{ij} = f(\xi, z, N) \tag{5.10}$$

where e_{ij} is the weight of the directed edge from country i to j. This is of course, a very simple model and apart from generating a network, it does not do much. Many important details are missing. For example, it is well known that trade volume is directly proportional to the product of GDPs and inversely proportional to the distance between a pair of countries. This feature is referred to as the gravity equation of trade (Barigozzi et al. 2010; Fagiolo 2010). Explaining the first part (trade being proportional to the product of GDPs) is not that difficult. Even in the model stated above a similar feature is embedded. The more difficult part is to understand why trade exactly inversely proportional (in the actual gravity equation, force of attraction is inversely proportional to the distance squared). Chaney (2014) presents a framework to understand that type of findings.

5.3.3 Migration Networks

Another important type of network is the migration network. People are moving across the world form country to country. An important motivation comes from productivity reasons which is related to wages or job-related reasons. The gravity equation kind of behavior also seen in migration as well. The network perspective in less prevalent in this literature even though there are instances of its usage (e.g. Stark and Jakubek 2013). Fagiolo and Mastrorillo (2014) connects the literature on trade and migration establishing that the trade network and the migration network are very correlated.

5.3.4 Financial Networks

A big part of the literature has focused on financial networks which broadly includes bank-to-bank transfers (Bech et al. 2010), firm-credit network (Bigio and Lao 2013), asset networks (Allen and Gale 2000; Babus and Allen 2009) etc. Jackson et al. (2014) proposes an extremely simple and abstract way to model interrelations between such entities. Suppose there are N primitive assets each with value r_n. There are organizations with cross-holding of claims. Thus the value of the i-th organization is

$$V_i = \sum_k w_{ik} r_k + \sum_k \tilde{w}_{ik} V_k \tag{5.11}$$

where w is a matrix containing relative weights of primitive asset holdings and $\tilde{w}$ is a matrix containing relative weights of cross-holding of claims. One can rewrite it in vector notations as

$$V = wr + \tilde{w}V \tag{5.12}$$

which can be rearranged to obtain

$$V = (I - \tilde{w})^{-1} wr. \tag{5.13}$$

One very interesting feature of this model is that

$$\sum_j V_j \geq \sum_j r_j. \tag{5.14}$$

The reason is that each unit of valuation held by one such organization contributes exactly 1 unit to its equity value, but at the same time through cross-holding of claims, it also increases value of other organizations. The net value is defined as

$$v_i = (1 - \sum_j \tilde{w}_{ji})V_i \tag{5.15}$$

which is simplified as

$$v = (I - w')(I - \tilde{w})^{-1} wr \tag{5.16}$$

with w' capturing the net valuation terms. Thus eventually this system also reduces to a recursive structure which we have already encountered in the input-output models. Finally they also introduce non-linearities and show how failures propagate through such a network. It is easily seen that any shock (even without any nonlinearities) to the primitive assets will affect all valuations through the cross-holding of claims channel. See Jackson et al. (2014) for details.

5.3.5 Dispersion on Networks: Kinetic Exchange Models

A descriptive model of inequality was forwarded by Drăgulescu and Yakovenko (2000) and Chakraborti and Chakrabarti (2000) (see Chakrabarti et al. 2013 for a general introduction and description of this class of models). The essential idea is that just like particles colliding against each other, people meet randomly in market places where they exchange money (the parallel being energy for particles). Thus following classical statistical mechanics, the the steady state distribution of money (energy) has an exponential feature which is also there in real world income and wealth distribution. Lack of rigorous micro-foundation is still a problem for such models, but the upside is that this basic model along with some variants of it can very quickly generate distributions of tradable commodities (money here) that resemble the actual distribution to a great extent including the power law tail. The equation describing evolution of assets (w) due to binary collision (between the i-th and the j-th agents) as

$$\begin{aligned} w_i(t+1) &= f(w_i(t), w_j(t), \varepsilon) \\ w_j(t+1) &= f(w_j(t), w_i(t), 1-\varepsilon) \end{aligned} \tag{5.17}$$

ε is a shock and $f(.)$ usually denotes a linear combination of its arguments Chakrabarti et al. (2013).

In principle, one can think of it as essentially a network model where agents are the nodes of the network and links randomly form among those agents and then destroyed. When the links are forms, they trade with each other and there is a microscopic transaction. After a sufficient number of transactions, the system reaches a steady state where the distribution of money does not change. Hence, inequality appears as a robust feature. An intriguing point is that the basic kinetic exchange models are presented in such a way that this underlying network structure is immaterial and does not have any effect on the final distribution. An open question is under what type of trading rule the topology of the network will have significant effect on the distribution (Chatterjee 2009). So far this issue has not been studied much.

5.3.6 *Networks and Growth*

So far all of the topics discussed are related to the idea of distribution and dispersion. Traditionally, economists have employed representative agent models to understand growth. In recent times, one interesting approach to understanding growth prospects has been proposed through applications of network theory. Hidalgo and Hausman (2009), suggested that existing complexity of an economy contains information about possible economic growth and development and they provide a measure of complexity of the production process using the trade network.

One unanswered point of the above approach is that it does not explain how such complexity is acquired and accumulated. A potential candidate is technology flow network. Theoretical attempts have mostly been concentrated on symmetric and linear technology flow networks (see e.g. Acemoglu 2008). An open modeling question is to incorporate non-trivial asymmetry in the technology network and model its impact on macroeconomic volatility.

5.3.7 *Correlation Networks*

This stream of literature is fundamentally different in its approach compared to the above. It is almost exclusively empirical with little theoretical (economic/finance) foundation. During the last two decades, physicists have shown interests in modeling stock market movements. The tools are obviously very different from the ones that economists use. In general, economists include stock market (if at all) in a macro model by assuming existence of one risky asset which typically corresponds to a market index. Sometimes multiple assets are considered but usually their interrelations are not explicitly modeled as the basic goal often is to study the risk-return trade off (hence one risky asset and one risk-free asset suffices in most cases). However, physicists took up exactly this problem: how to model joint evolution and interdependence of a large number of stocks? Plerou et al. (2002) introduced some important tools.

Later extensions can be found in Bonanno et al. (2003), Tumminello et al. (2010) and references therein. These studies were done in the context of developed countries with little attention to less developed financial markets. Pan and Sinha (2007) extended the study on that frontier.

The basic object of study here is the correlation network between N number of stocks. Each stock generates a return $\{r_{nt}\}$ of length T. Given any two stocks i and j, one can compute the correlation matrix with the i, j-th element

$$c_{ij} = \frac{E(r_i \cdot r_j) - E(r_i) \cdot E(r_j)}{\sigma_i \cdot \sigma_j}. \tag{5.18}$$

Clearly the correlation matrix is symmetric. Then the question is if there is any way one can divide the correlation matrix into separate modes defining the market (m), group (g) and idiosyncratic (r) effects,

$$C = C^m + C^g + C^r. \tag{5.19}$$

A very important tool is provided by the random matrix theory which allows us to pin down the idiosyncratic effects. Through eigenvalue decomposition, one can construct the random mode by the so-called Wishart matrix which is essentially a random correlation matrix (Pan and Sinha 2007). This helps to filter the random component of the correlation matrix (C^r in Eq. 5.19). The market mode is the global factor that affects all stocks simultaneously in the same direction. The mode in between these two, is described as the group mode.

Empirically, the group effects in certain cases seems to be very prominent and correspond to actual sector-specific stocks (Plerou et al. 2002), but not always (Kuyyamudi et al. 2015; Pan and Sinha 2007). Onnela et al. (2003) also studied contraction and expansion along with other properties of the correlation network. However, this field is still not very matured and the applications of such techniques are not widespread.

One can also construct networks based on the correlation matrix. A standard way to construct a network would be to consider only the group correlation matrix and apply a threshold to determine if stocks are connected or not. A complementary approach without using a threshold would be to construct a metric based on the correlation coefficients. This is very useful to construct a minimum spanning tree and study its evolution over time, (see Bouchaud and Potters (2009) for related discussions).

5.4 Summary

One recurring theme in all economic models of network is that the building blocks are countries or firms or cities and usually one comes up with an input-output structure defined over those economic entities, on the variable of interest. The input-output network is essentially a flow network (of people, commodity or money). Very early attempts to build these models were based on market clearing by fixing quantities as if the economy is run by a central planner. The newer versions have explicit utility maximization and cost minimization. Thus most of these models use general equilibrium theory to solve for the final allocation. There are many game-theoretic problems as well which use network terminology and tools, which we ignore the present purpose. See Jackson (2010) for a detailed review.

This indicates a possibility that general equilibrium theory on networks can provide a middle ground between standard macroeconomics where granularity of agents do not mater and agent-based models where granularity matters but is so much dependent on the model specification that there is no consistent theory. In particular this might proves useful for explaining economic fluctuations (La'o 2014).

In this short note, a number of applications of network theory have been presented to understand macroeconomic patterns. Some open and unsettled problems in the theories have also been discussed. Extensions of the previous work and further explorations on the deeper relationships between network topology and macro behavior will be useful, probably more useful than many other well established branches in modern economics.

References

D. Acemoglu, *Introduction to modern economic growth*. Princeton University Press, 2008.

D. Acemoglu, V. Carvalho, A. Ozdaglar, and A. Tahbaz-Salehi, "The network origin of economic fluctuations," *Econometrica*, vol. 80, pp. 1977–2016, 2012.

R. Albert and A.-L. Barabasi, "Statistical mechanics of complex networks," *Review of modern physics*, vol. 74, pp. 47–97, 2002.

F. Allen and D. Gale, "Financial contagion," *Journal of political economy*, vol. 108, no. 1, pp. 1–33, 2000.

A. Babus and F. Allen, "Networks in finance," in *Network-based Strategies and Competencies* (P. Kleindorfer and J. Wind, eds.), pp. 367–382, 2009.

P. Bak, K. Chen, J. Scheinkman, and m. Woodford, "Aggregate fluctuations from independent sectoral shocks: self-organized criticality in a model of production and inventory dynamics," *Ricerche Economiche*, vol. 47-1, 1993.

M. Barigozzi, G. Fagiolo, and D. Garlaschelli, "Multinetwork of international trade: A commodity-specific analysis," *Physical Review E*, vol. 81, no. 4, p. 046104, 2010.

A. Barrat, M. Barthelemy, and A. Vespignani, *Dynamical Processes on Complex Networks*. Cambridge University Press, 2008.

M. Bech, J. Chapman, and R. Garrat, "Which bank is the "central" bank?," *Journal of monetary economics*, vol. 57, pp. 352–363, 2010.

S. Bigio and J. Lao, "Financial frictions in production networks." 2013.

G. Bonanno, G. Caldarelli, F. Lillo, and R. N. Mantegna, "Topology of correlation-based minimal spanning trees in real and model markets," *Physical Review E*, vol. 68, no. 4, p. 046130, 2003.

J.-P. Bouchaud and M. Potters, *Theory of Financial Risk and Derivative Pricing*. CUP, 2009.

M. Buchanan, "To understand finance, embrace complexity." http://www.bloomberg.com/news/2013-03-10/to-understand-finance-embrace-complexity.html, March 11, 2013.

G. Caldarelli, *Scale-free networks: complex webs in nature and technology*. Oxford Univ. Press, UK., 2007.

B. K. Chakrabarti, A. Chakraborti, S. R. Chakravarty, and A. Chatterjee, *Econophysics of income and wealth distributions*. Cambridge Univ. Press, Cambridge, 2013.

A. Chakraborti and B. K. Chakrabarti, "Statistical mechanics of money: how saving propensity affects its distribution," *Eur. Phys. J. B*, vol. 17, pp. 167–170, 2000.

T. Chaney, "The network structure of international trade," *American Economic Review*, vol. (forthcoming), 2014.
A. Chatterjee, "Kinetic models for wealth exchange on directed networks," *Eur. Phys. J. B*, vol. 67, no. 4, pp. 593–598, 2009.
S. N. Dorogovtsev and J. F. F. Mendes, *Evolution of Networks: From Bilogical Nets to the Internet and WWW*. Oxford University Press, 2003.
A. Drăgulescu and V. Yakovenko, "Statistical mechanics of money," *Eur. Phys. J. B*, vol. 17, pp. 723–729, 2000.
W. Dupor, "Aggregation and irrelevance in multi-sector models," *Journal of Monetary Economics*, vol. 43, p. 391, 1999.
P. Erdos and A. Renyi, "On random graphs i," *Publicationes Mathematicae*, vol. 6, pp. 290–297, 1959.
G. Fagiolo, "The international-trade network: gravity equations and topological properties," *Journal of Economic Interaction and Coordination*, vol. 5, no. 1, pp. 1–25, 2010.
G. Fagiolo and M. Mastrorillo, "Does human migration affect international trade? a complex-network perspective," *Plos One*, vol. 9, no. 5, p. e97331, 2014.
A. T. Foerster, P. G. Sarte, and M. W. Watson, "Sectoral vs. aggregate shocks: A structural factor analysis of industrial production," *Journal of Political Economy*, vol. 119(1), pp. 1–38, 2011.
C. Hidalgo and R. Hausman, "The building blocks of economic complexity," *Proceedings of national academy of sciences*, vol. 106 (25), pp. 10570–10575, 2009.
M. Horvath, "Cyclicality and sectoral linkages: Aggregate fluctuations from independent sectoral shocks," *Review of Economic Dynamics*, vol. 1, pp. 781–808, 1998.
M. Jackson, *Social and Economic Networks*. Princeton University Press, 2010.
M. Jackson, M. Elliott, and B. Golub, "Financial networks and contagion," *American Economic Review*, vol. 104, no. 10, pp. 3115–53, 2014.
P. Krugman, *The Self-Organizing Economy*. Wiley-Blackwell, 1996.
C. Kuyyamudi, A. S. Chakrabarti, and S. Sinha, "Long-term evolution of the topological structure of interactions among stocks in the new york stock exchange 19252012," in *Econophysics and Data Driven Modelling of Market Dynamics*, 2015.
J. La'o, "A traffic jam theory of recessions." 2014.
W. Leontief, *Input-Output Economics (2nd Ed.)*. Oxford University Press (New York), 1986.
W. Leontief, "Quantitative input and output relations in the economic system of the united states," *Review of Economics and Statistics*, vol. 18, pp. 105–125, 1936.
W. Leontief, "Structural matrices of national economies," *Econometrica*, vol. 17, pp. 273–282, 1947.
J. P. Onnela, A. Chakraborti, K. Kaski, J. Kertesz, and A. Kanto, "Dynamics of market correlations: Taxonomy and portfolio analysis," *Physical Review E*, vol. 68, no. 056110, 2003.
R. Pan and S. Sinha, "Collective behavior of stock price movements in an emerging market," *Physical Review E*, vol. 76, no. 046116, pp. 1–9, 2007.
V. Plerou, P. Gopikrishnan, B. Rosenow, L. A. N. Amaral, Guhr, and H. E. Stanley, "Random matrix approach to cross correlations in financial data," *PHYSICAL REVIEW E*, vol. 65, no. 066126, 2002.
F. Schweitzer, G. Fagiolo, D. Sornette, F. Vega-Redondo, A. Vespignani, and D. R. White, "Economic networks: The new challenges," *Science*, vol. 325, no. 5939, p. 422, 2009.
O. Stark and M. Jakubek, "Migration networks as a response to financial constraints: onset and endogenous dynamics," *Journal of Development Economics*, vol. 101, pp. 1–7, 2013.
M. Tumminello, F. Lillo, and R. N. Mantegna, "Correlation, hierarchies, and networks in financial markets," *Journal of Economic Behavior & Organization*, vol. 75, no. 1, pp. 40–58, 2010.
D. J. Watts and S. H. Strogatz, "Collective dynamics of 'small-world' networks," *Nature*, vol. 393, no. 6684, pp. 440–442, 1998.

Chapter 6
Power Law Distributions for Share Price and Financial Indicators: Analysis at the Regional Level

Michiko Miyano and Taisei Kaizoji

Abstract We investigate whether the distribution of share price follows a power law distribution at the regional level, using data from companies publicly listed worldwide. Based on ISO country codes, 7,796 companies are divided into four regions: America, Asia, Europe, and the rest of the world. We find that, at the regional level, the distributions of share price follow a power law distribution and that the power law exponents estimated by region are quite diverse. The power law exponent for Europe is close to that of the world and indicates a Zipf distribution. We also find that the theoretical share price and fundamentals estimated using a panel regression model hold to a power law at the regional level. A panel regression in which share price is the dependent variable and dividends per share, cash flow per share, and book value per share are explanatory variables identifies the two-way fixed effects model as the best model for all regions. The results of this research are consistent with our previous findings that a power law for share price holds at the world level based on panel data for the period 2004–2013 as well as cross-sectional data for these 10 years.

6.1 Introduction

Since Vilfredo Pareto (1848–1923) found more than 100 years ago that income distributions follow a power law, numerous studies have attempted to find and explain this phenomenon using a variety of real world data. Power laws, including Zipf's law, appear widely in physics, biology, economics, finance, and the social sciences. Newman (2005) introduced examples of distributions that appear to follow power laws in a variety of systems, including Word frequency, Citations of scientific papers,

M. Miyano · T. Kaizoji (✉)
Graduate School of Arts Sciences, International Christian University,
3-10-2 Osawa, Mitaka, Tokyo 181-8585, Japan
e-mail: kaizoji@icu.ac.jp

M. Miyano
e-mail: g199006t@icu.ac.jp

F. Abergel et al. (eds.), *Econophysics and Sociophysics: Recent Progress and Future Directions*, New Economic Windows,
DOI 10.1007/978-3-319-47705-3_6

Web hits, Copies of books sold, Telephone calls, Magnitude of earthquakes, Diameter of moon craters, Intensity of solar flares, Intensity of wars, Wealth of the richest people, Frequency of family names, and Populations of cities. All have been proposed to follow power laws by researchers. Examples of Zipf's law, whose power exponent is equal to one, are also found in a variety of systems. Gabaix (1999) showed that the distribution of city sizes follows Zipf's law. Axtell (2001) showed that distributions of company sizes follow Zipf's law. Econophysics focuses on the study of power laws in economies and financial markets. (For a recent review of the development of Econophysics, see Chakraborti et al. (2011a, b)).

Using cross-sectional data for the period 2004–2013 from companies publicly listed worldwide, Kaizoji and Miyano (2016a) showed that share price and financial indicators per share follow a power law distribution. For each of the 10 years examined, a power law distribution for share price was verified. Using panel data for the same period, Kaizoji and Miyano (2016b) developed an econometric model for share price and showed that a two-way fixed effects model identified from a panel regression with share price as the dependent variable and dividends per share, cash flow per share, and book value per share as explanatory variables effectively explains share price. Based on the same data, Kaizoji and Miyano (2016c) also found that share price and certain financial indicators per share follow Zipf's law and verified that company fundamentals estimated using a two-way fixed effects model also follow Zipf's law.

The aim of this current study is to (1) verify that the distributions of share price and fundamentals follow power laws at the regional level, and (2) investigate the regional characteristics of share price behavior following previous studies (Kaizoji and Miyano 2016a, b, c).

For this study, a number of companies listed worldwide were divided into four regions: America, Asia, Europe, and rest of the world.[1]

Using this scheme, we found that the distributions of share price and financial indicators per share follow a power law distribution at the regional level. Further, we found that the distribution of fundamentals estimated using the two-way fixed effects model that was selected as the best model for all regions follows a power law distribution at the regional level and that the estimated power law exponents for share price are quite diverse by region, in the range 0.98–3.42.

This paper is organized as follows: Section 6.2 gives an overview of the share price data at the regional level; Section 6.3 describes the econometric model and presents the estimated results; Section 6.4 examines the estimated distributions of the fundamentals; Section 6.5 examines the financial indicators per share data used in the study; Section 6.6 concludes.

[1]America includes North America, South America, and Central America. Asia includes eastern Asia, southern Asia, central Asia, and the Middle East. The rest of the world includes Oceania and Africa.

6.2 Data

The data source used here is the OSIRIS database provided by Bureau Van Dijk containing financial information on globally listed companies. In this study, we employ annual data for the period 2004–2013. Stock and financial data for a total of 7,796 companies for which data were available over this 10-year period were extracted from the database. Using this data, we performed a statistical investigation of share price and dividends per share, cash flow per share, and book value per share, all of which were obtained by dividing available values by the number of shares outstanding.

For analysis at the regional level, we divided the 7,796 companies selected into the four regions described above, using the ISO country code appropriate to the individual companies. The number of companies in each region was as follows: America, 1,886 companies; Asia, 4,065 companies; Europe, 1,436 companies; the rest of the world, 409 companies.[2]

6.2.1 Power Law Distributions of Share Price in Regional Data

Using the same company data in a previous study, Kaizoji and Miyano (2016c) found that in the upper tail, which includes approximately 2 % of the total observations, the distributions of share price and financial indicators per share follow a power law distribution at the worldwide level. In this section, we investigate whether the distributions of share price follow a power law distribution at the regional level.

Defining a power law distribution is straightforward. Let x represent the quantity in whose distribution we are interested. (In our research, x represents share price or various financial indicators per share.)

Observed variable, X follows a power law distribution if its distribution is described by[3]:

$$Pr(X > x) = 1 - F(x) = (\frac{x}{k})^{-\alpha}, \quad x \geq k > 0 \tag{6.1}$$

where $F(x)$ denotes the cumulative distribution function, k denotes a scaling parameter corresponding to the minimum value of the distribution, and α denotes the shape parameter. We call α power law exponent.

By taking the logarithm of both sides of Eq. (6.1), the following equation is obtained:

$$ln(Pr(X > x)) = \alpha lnk - \alpha lnx \tag{6.2}$$

[2]Total observations available in each region were as follows: America, 8935; Asia, 27,407; Europe, 8,791; rest of the world, 2028.

[3]The probability density function for the Pareto distribution is defined as $f(x) = \frac{\alpha k^{\alpha}}{x^{\alpha+1}}, \quad x \geq k > 0.$

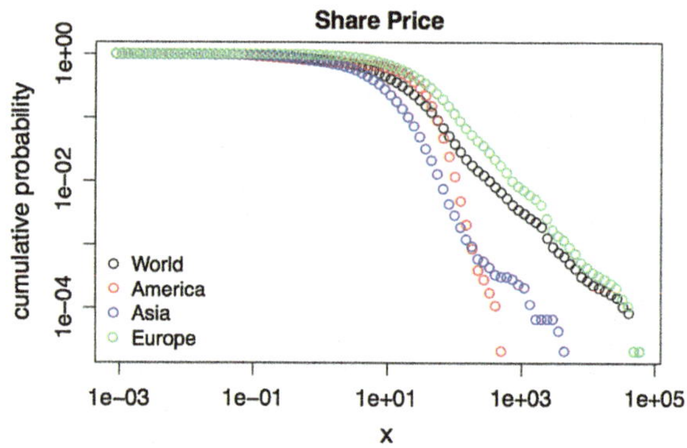

Fig. 6.1 The complementary cumulative distribution of share price (log-log plot)

If the distribution is plotted using logarithmic horizontal and vertical axes and appears approximately linear in the upper tail, we can surmise that the distribution follows a power law distribution.

As a first step, we plotted complementary cumulative distributions for the data. Figure 6.1 shows the complementary cumulative distributions of share price by region, with logarithmic horizontal and vertical axes.[4]

From Fig. 6.1, the regional complementary cumulative distributions of share price seem to be roughly linear in their upper tails, although the slopes differ, suggesting that the distributions of share price follow a power law distribution at the regional level. In addition, the distribution of share price for Europe appears to be close to that of the world in the upper tail. This suggests that a power law distribution for the world mostly originates from the European data.

In the second step, we estimate the power law exponent using the MLE (maximum likelihood estimator) method. The MLE is given by

$$\hat{\alpha} = n[\sum_{i=1}^{n} ln(\frac{x_i}{x_{min}})]^{-1} \tag{6.3}$$

where $\hat{\alpha}$ denotes the estimates of α, $x_i, i = 1, \cdots, n$ are observed values of x, and $x_i > x_{mim}$.[5] The results are presented in Table 6.1.

To test whether the distributions of share price observed at the regional level follow a power law distribution, we use a Cramér-von Mises test, one of the goodness-of-fit tests based on a measurement of distance between the distribution of empirical data and the hypothesized model. The distance is usually measured either by a supremum or a quadratic norm. The Kolmogorov-Smirnov statistic is a well-known supremum norm. The Cramér-von Mises family, using a quadratic norm, is given by

[4]The graph fo the rest of the world is excluded. This is done throughout since the numbers of companies in this is only 5.2 % of the total.

[5]Details of the derivations are presented (Kaizoji and Miyano 2016c).

Table 6.1 Estimates of power law exponents and p-values calculated in the test for a power law distribution

Region	Power law exponents	xmin	Cramér-von Mises test p-value	Tail %	Total observations
World	1.003	133.4	0.132	2.0	47,161
America	3.427	75.1	0.676	3.0	8,935
Asia	2.096	23.1	0.099	6.5	27,407
Europe	0.975	115.5	0.169	11.0	8,791
Rest of the world	1.220	97.9	0.162	2.4	2,028

$$Q = n \int_{-\infty}^{\infty} [F_n(x) - F(x)]^2 \psi(x) dF(x) \tag{6.4}$$

where $F_n(x)$ denotes the empirical distribution function and $F(x)$ denotes the theoretical distribution function. When $\psi(x) = 1$, Q is the Cramér-von Mises statistic, denoted W^2. When $\psi(x) = [F(x)(1 - F(x))]^{-1}$, Q is the Anderson and Darling statistic, denoted A^2.

Although a variety of goodness-of-fit statistics have been proposed, we use the W^2 statistic of Cramér-von Mises in our research.[6] The test statistic of the Cramér-von Mises test is given by Eq. (6.4), with $\psi(x) = 1$.[7]

The null hypothesis is that the distribution of observed data follows a power law distribution. Table 6.1 presents the estimates of the power law exponents and p-values for the tests.

As is evident here, the null hypothesis cannot be rejected at the 5 % significant level for all regions. As can be seen, the power law exponents are quite diverse by region. The power law exponent for Europe is close to that of the world, while the power law exponents for America and Asia are 3.4 and 2.1, respectively.

6.2.2 *Changes in Averages and Variances of Share Price*

Figures 6.2 and 6.3 show the changes in the averages and variances of share price, respectively. As indicated, Europe shows a notably high average, while Asia shows the lowest. All regions appear to show the same pattern of averages, experiencing an apparent fall in 2008 and slowly recovering after 2009. Regarding the changes in variance, in contrast with the average, America shows the largest variance and Europe shows the smallest variance before 2009. Variances tend to decline slightly after 2007, except for Europe, where they tend to rise slightly.

---

[6] According to D'Agostino and Stephans (1986), the Anderson and Darling test is in common use. However, the test is found to be highly conservative by Clauset et al. (2009)

[7] The two classes of measurement and computational details for this test are found in Čížek and Weron (2005, Chap. 13)

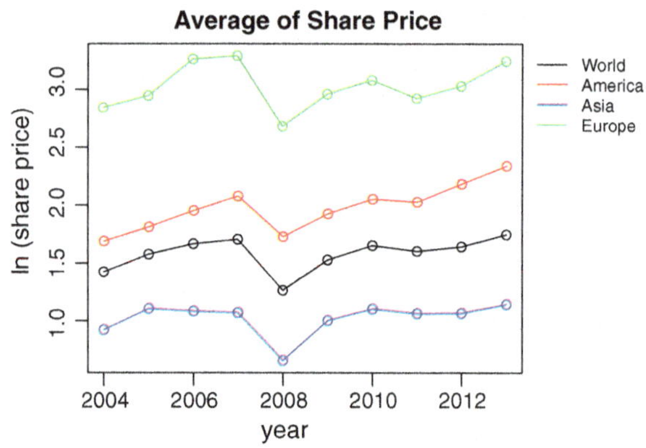

Fig. 6.2 Changes in average of share price

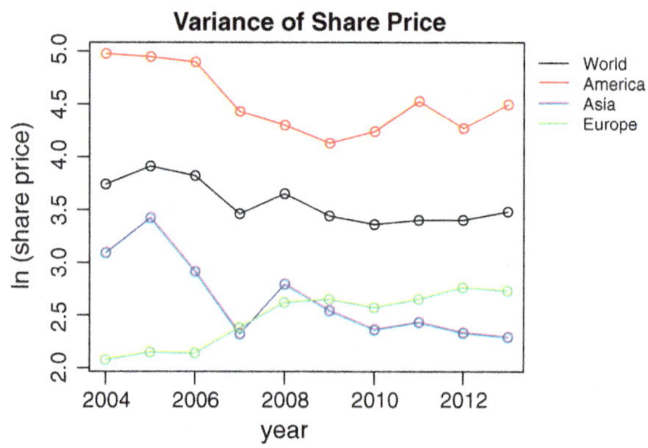

Fig. 6.3 Changes in variance of share price

6.3 Econometric Model for Fundamentals

In our previous research, the model we chose to use was shown to have quite a high explanatory power with respect to share price. Therefore, we used a similar econometric model for our regional data in this current study.

6.3.1 Econometric Model

Assuming the relationship between share price and the set of financial indicators, that includes dividends per share, cash flow per share, and book value per share, to be logarithmic linear, the econometric model for our study can be written as

$$lnY_{it} = lna + b_1 lnX_{1,it} + b_2 lnX_{2,it} + b_3 lnX_{3,it} + u_{it} \quad i = 1, \cdots, N; \quad t = 1, \cdots, T \tag{6.5}$$

where i denotes cross-section (i.e., individual company), t denotes time point (year), and

Y_{it}: share price for company i, at year t
a: constant term
$X_{1,it}$: dividends per share for company i, at year t
$X_{2,it}$: cash flow per share for company i, at year t
$X_{3,it}$: book value per share for company i, at year t
u_{it}: error term

We estimate the model in Eq. (6.5) using the Panel Least Squares method. In the panel regression model, the error term u_{it} can be assumed to be the two-way error component model. Details of the two-way error component model are described in Kaizoji and Miyano (2016b). The estimation models examined include the pool OLS model, the individual fixed effects model, the time fixed effects model, the two-way fixed effects model, the individual random effects model, and the time random effects model.[8]

We perform the estimation by region. That is, we estimate 4×6 models using the same method as the world model. The model selection tests are as follows: the likelihood ratio test and F-Test for the selection of the pool OLS model versus the fixed effects model; and the Hausman test for the selection of the random effects model versus the fixed effects model. The selection test for the pool OLS model vs the random effects model is based on the simple test proposed by Woodlridge (2010).[9]

For all regions, the two-way fixed effects model is identified as the best model among the six alternatives. In a two-way fixed effects model, the error term consists of the following three terms:

$$u_{it} = \mu_i + \gamma_t + \varepsilon_{it} \tag{6.6}$$

μ_i: unobservable individual fixed effects
γ_t: unobservable time fixed effects
ε_{it}: pure disturbance

μ_i is the individual fixed effect and represents the company's effect on share price (among other factors). γ_t is the time fixed effect and is related to the point in time (year) affecting stock markets, among other factors (for example, factors caused by financial and economic shocks such as Global financial crisis in 2008).

Table 6.2 shows the regional results for the panel regression model described in Eq. (6.5). The signs of the three coefficients are all positive, consistent with corporate value theory. The p-values of the coefficients are quite small, indicating statistical significance for all regions. In addition, the R^2 values are in the range 0.95–0.98, indicating that the estimated models explain the variation in share prices quite well.

[8]The two-way random effects model cannot be used since we use unbalanced panel observations.

[9]Woodlridge (2010, p.299) proposes a method that uses residuals from pool OLS and checks the existence of serial correlations.

Table 6.2 Regional results of panel regression (two-way fixed effects model). Total observations presented in the table are unbalanced panel observations

Region		lna	b_1	b_2	b_3	R^2	Total observations
World	Coefficient	1.485	0.137	0.208	0.378	0.969	47,161
	Std. error	0.014	0.003	0.004	0.007		
	p-value	0.000	0.000	0.000	0.000		
America	Coefficient	1.750	0.119	0.215	0.324	0.977	8,935
	Std. error	0.024	0.007	0.009	0.013		
	p-value	0.000	0.000	0.000	0.000		
Asia	Coefficient	1.154	0.112	0.218	0.440	0.956	27,404
	Std. error	0.020	0.005	0.005	0.011		
	p-value	0.000	0.000	0.000	0.000		
Europe	Coefficient	2.160	0.191	0.154	0.318	0.967	8,791
	Std. error	0.035	0.007	0.008	0.014		
	p-value	0.000	0.000	0.000	0.000		
Rest of the world	Coefficient	1.546	0.189	0.207	0.416	0.953	2,028
	Std. error	0.038	0.014	0.019	0.027		
	p-value	0.000	0.000	0.000	0.000		

The econometric model for share price, using dividends per share, cash flow per share, and book value per share as explanatory variables, fits the actual data quite well at the regional level as well as at the world level.

Among the three financial indicators, the coefficient of book value per share (b_3) is largest in all regions, while the coefficient of dividends per share (b_1) is smallest, except for Europe. The constant term for Europe is quite large compared to the other regions.

6.3.2 *Theoretical Value and Fundamentals*

By multiplying both sides of Eq. (6.5) by the exponent function, $\hat{Y}$is obtained as written:

$$\hat{Y} = \hat{a}(X_1^{\hat{b}_1} X_2^{\hat{b}_2} X_3^{\hat{b}_3})(e^{\hat{\mu}_i})(e^{\hat{\gamma}_t}) \tag{6.7}$$

where $\hat{Y}$ is the estimated value for share price, which we call the theoretical value.

We can remove the time fixed effects term, γ_t, from the error term described in (6.6). After subtracting the time effects term from Eq. (6.6), $\tilde{Y}$is obtained by multiplying both sides of Eq. (6.5) by the exponent function:

$$\tilde{Y} = \hat{a}(X_1^{\hat{b}_1} X_2^{\hat{b}_2} X_3^{\hat{b}_3})(e^{\hat{\mu}_i}) \tag{6.8}$$

Table 6.3 The results of two-sample Kolmogorov-Smirnov test and correlation of theoretical value and fundamentals with share price

Region		K-statistic	p-value	Correlation coefficient
World	Theoretical value	0.006	0.287	0.984
	Fundamentals	0.007	0.253	0.982
America	Theoretical value	0.016	0.228	0.988
	Fundamentals	0.018	0.121	0.986
Asia	Theoretical value	0.009	0.215	0.987
	Fundamentals	0.013	0.023	0.974
Europe	Theoretical value	0.010	0.757	0.983
	Fundamentals	0.014	0.384	0.977
Rest of the world	Theoretical value	0.016	0.950	0.976
	Fundamentals	0.016	0.950	0.971

As in our previous studies (Kaizoji and Miyano 2016b, c), we identify $\tilde{Y}$ as the company fundamentals since the time effect common to all companies has been removed, leaving only the company fundamentals.

We investigated the distribution of fundamentals in the upper distribution tail by region. As described in Sect. 6.2, the distribution of share price follows a power law distribution at the regional level. Before investigating the distribution of fundamentals, we examined whether the distribution of fundamentals coincided with that of share price. Using a two-sample Kolmogorov-Smirnov test, we tested goodness-of-fit between company fundamentals and share price. Table 6.3 shows the results of the test as well as the relevant correlation coefficients. Given the test results shown in Table 6.3, the null hypothesis that the two distributions coincide cannot be rejected at the 5 % significant level, except in the case of Asia. With respect to the theoretical value, the null hypothesis cannot be rejected the 5 % significant level for all regions. Correlation coefficients with share price are in the range 0.97–0.99.

6.4 Power Law Distribution for Fundamentals

The complementary cumulative distributions of theoretical value and company fundamentals are shown in Figs. 6.4 and 6.5. As described in the previous section, the theoretical value is directly estimated using a two-way fixed effects model, while the fundamentals are computed by removing the time fixed effects term from the theoretical value. The two figures are plotted with logarithmic horizontal and vertical axes. Both figures show that the upper tails of the distributions appear roughly linear, although there are small differences among the regions. The distributions in Figs. 6.4 and 6.5 are quite similar to those shown in Fig. 6.1 for share price, suggesting that the regional distributions of theoretical value and company fundamentals also follow a power law distribution.

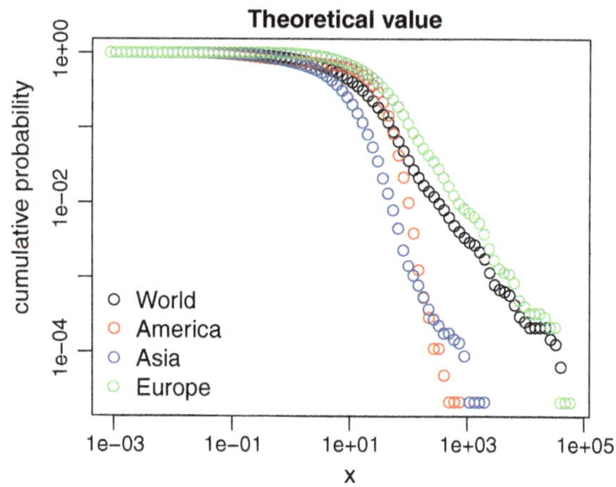

Fig. 6.4 The complementary cumulative distribution of theoretical value (log-log plot)

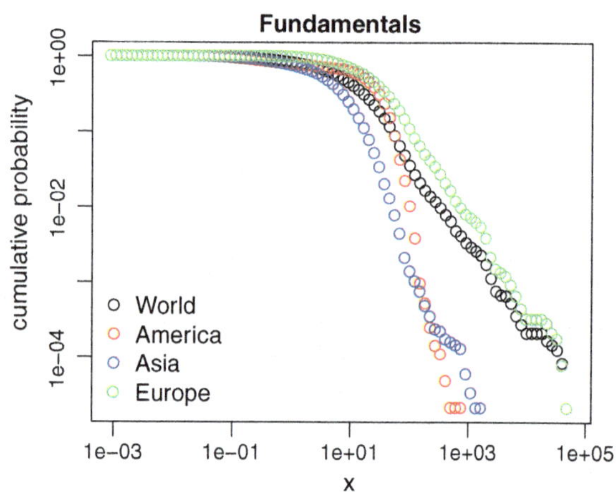

Fig. 6.5 The complementary cumulative distribution of fundamentals (log-log plot)

As described in Sect. 6.2.1, we performed an estimation of the power law exponents using the MLE method, then tested the null hypothesis of a power law distribution using the Cramér-von Mises test. Table 6.4 shows the estimated power law exponents and test results for the power law distribution hypothesis. For easy comparison with share price, we show the power law exponents and p-values for share price that were shown earlier in Table 6.1.

The power law exponents for theoretical value are similar to those of the company fundamentals and are close to share price, except in the case of America. For America, the power law exponents are extremely large for theoretical value and fundamentals, and differ from that of share price. The Cramér-von Mises test fails to reject the null hypothesis of a power law distribution for all regions. From these results, it can be said that the distributions of theoretical value and fundamentals follow a power law distribution at the regional level.

Table 6.4 Estimates of power law exponents and p-values calculated in the test for a power law distribution

Region		Power law exponents	xmin	Cramér-von Mises test p-value	Tail %	Total observations
World	Share price	1.003	133.4	0.132	2.0	47,161
	Theoretical value	1.012	128.5	0.106	2.0	
	Fundamentals	1.006	119.7	0.128	2.1	
America	Share price	3.427	75.1	0.676	3.0	8,935
	Theoretical value	3.814	72.7	0.119	3.0	
	Fundamentals	3.993	72.8	0.115	3.1	
Asia	Share price	2.096	23.1	0.099	6.5	27,407
	Theoretical value	2.172	27.4	0.106	4.0	
	Fundamentals	2.183	25.5	0.089	4.5	
Europe	Share price	0.975	115.5	0.169	11.0	8,791
	Theoretical value	0.975	123.7	0.170	10.0	
	Fundamentals	0.963	108.5	0.156	11.0	
Rest of the world	Share price	1.220	97.9	0.162	2.4	2,028
	Theoretical value	1.238	84.5	0.080	3.0	
	Fundamentals	1.251	86.4	0.123	3.0	

As can be seen here, the power law exponents of theoretical value and company fundamentals for Europe are close to 1, as are those for the world.

6.5 Power Law Distribution for Financial Indicators per Share

In the previous section, we showed that the distribution of fundamentals follows a power law distribution at the regional level. Kaizoji and Miyano (2016c) suggested that the reason why the distribution of company fundamentals follows a power law distribution is due to the fact that the distributions of financial indicators per share, representing corporate value, follows a power law distribution. In this study, we examined whether the distributions of financial indicators per share follow a power law distribution at the regional level.

Figures 6.6, 6.7 and 6.8 show the complementary cumulative distributions of dividends per share, cash flow per share, and book value per share using logarithmic horizontal and vertical axes. In these figures, it appears that the regional complementary cumulative distributions of financial indicators per share are roughly linear in their upper tails.

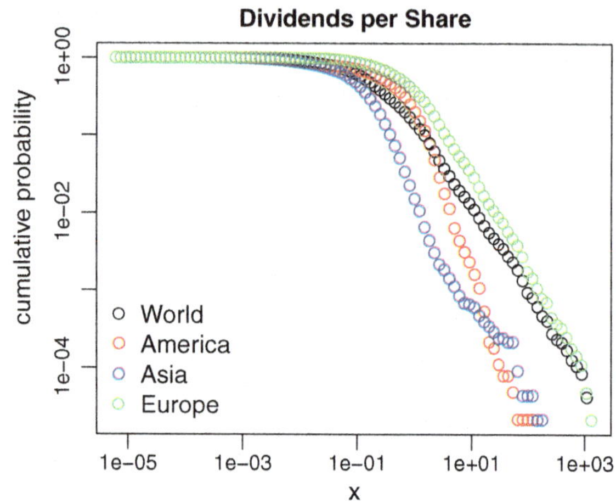

Fig. 6.6 The complementary cumulative distribution of dividends per share (log-log plot)

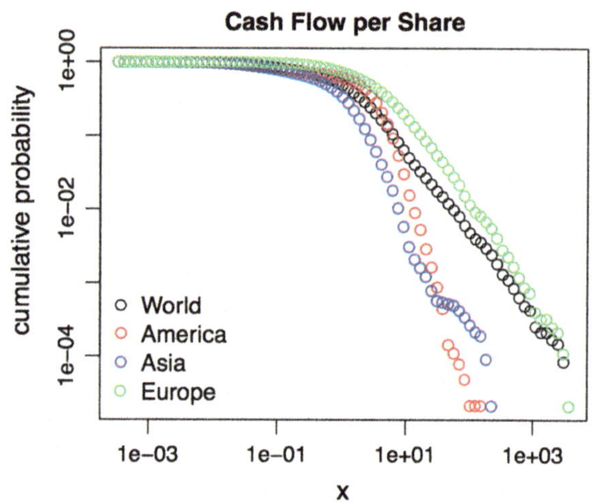

Fig. 6.7 The complementary cumulative distribution of cash flow per share (log-log plot)

We show the estimated power law exponents and test results of the power law hypothesis in Table 6.5. Tests for the distributions of dividends per share and cash flow per share fails to reject the null hypothesis at 5 % significant level for America, Europe, and the rest of the world. However, the null hypothesis is rejected for Asia. For book value per share, the null hypothesis cannot be rejected at the 5 % significant level for all regions.

Figures 6.9, 6.10 and 6.11 show the changes in average of dividends per share, cash flow per share, and book value per share. The changes in average for dividends per share and cash flow per share fell slightly in 2009 except Asia. However, changes in average book value per share do not appear clearly in 2008–2009. As shown in Fig. 6.2, share prices in all regions seem to have been immediately affected by the

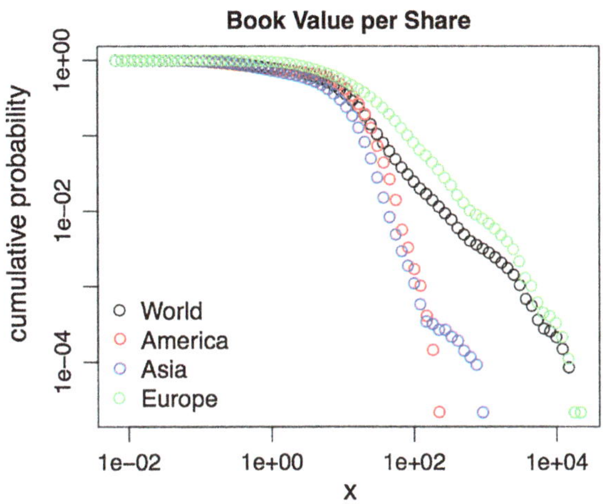

Fig. 6.8 The complementary cumulative distribution of book value per share (log-log plot)

Table 6.5 Estimates of power law exponents and p-values calculated in the test for a power law distribution

Region	Financial indicator per share	Power law exponents	xmin	Cramér-von Mises test p-value	Tail %	Total observations
World	Dividends	1.015	3.6	0.130	2.6	47,161
	Cash flow	1.051	21.9	0.151	2.0	
	Book value	0.955	98.9	0.221	2.0	
America	Dividends	2.029	1.6	0.361	10.0	8,935
	Cash flow	2.515	5.6	0.798	10.0	
	Book value	2.918	58.1	0.363	1.1	
Asia	Dividends	1.757	0.5	0.003	4.0	27,407
	Cash flow	2.138	3.6	0.001	4.0	
	Book value	2.261	22.7	0.082	4.0	
Europe	Dividends	0.956	2.8	0.123	14.2	8,791
	Cash flow	1.017	18.7	0.071	11.2	
	Book value	0.929	86.6	0.084	11.3	
Rest of the world	Dividends	1.074	4.0	0.063	2.5	2,028
	Cash flow	1.106	13.4	0.299	2.5	
	Book value	1.282	35.3	0.188	3.0	

global crisis in 2008, while the impact on the financial indicators per share seems to appear one year later.

Figures 6.12, 6.13 and 6.14 show the changes in the variance of dividends per share, cash flow per share, and book value per share. In contrast with the average, the variances of share price and the financial indicators per share are largest for America

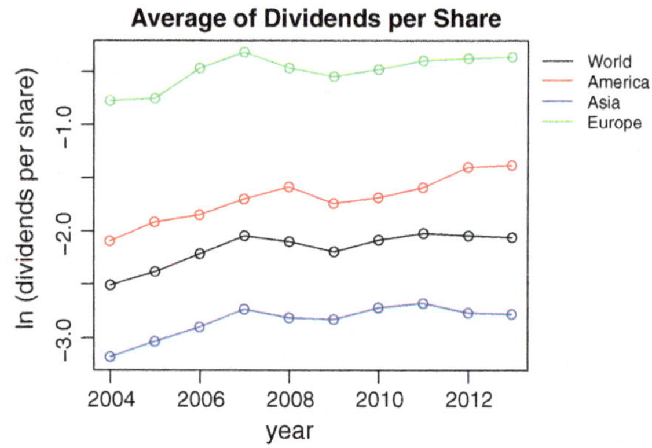

Fig. 6.9 Changes in average of dividends per share

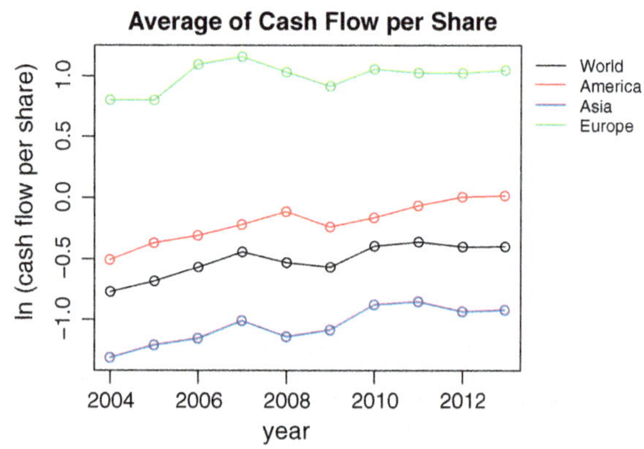

Fig. 6.10 Changes in average of cash flow per share

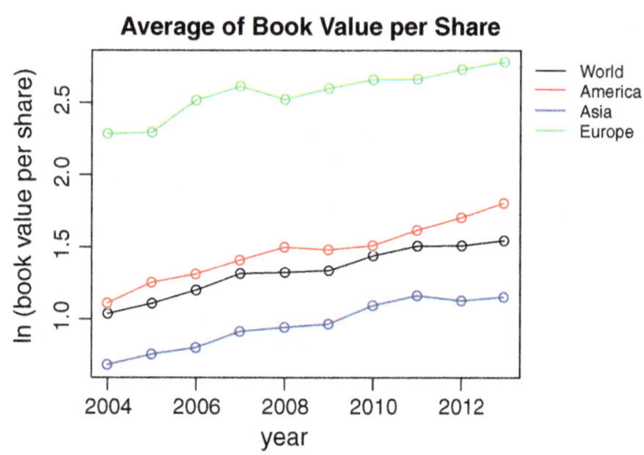

Fig. 6.11 Changes in average of book value per share

and relatively small for Europe. In addition, the variances of book value per share for America show a sharp decline from 2006 to 2008 and a relatively small difference from the other regions after 2008.

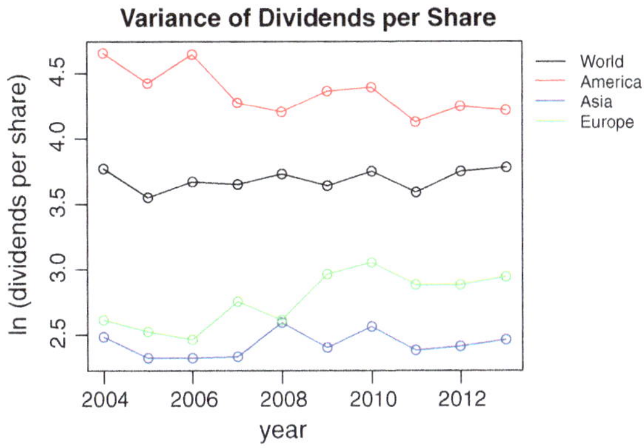

Fig. 6.12 Changes in variance of dividends per share

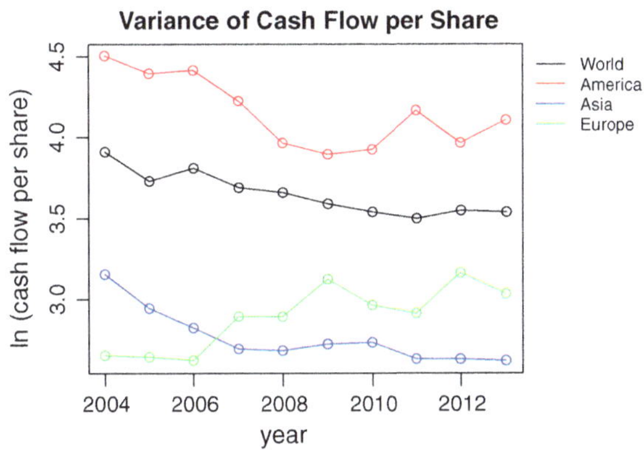

Fig. 6.13 Changes in variance of cash flow per share

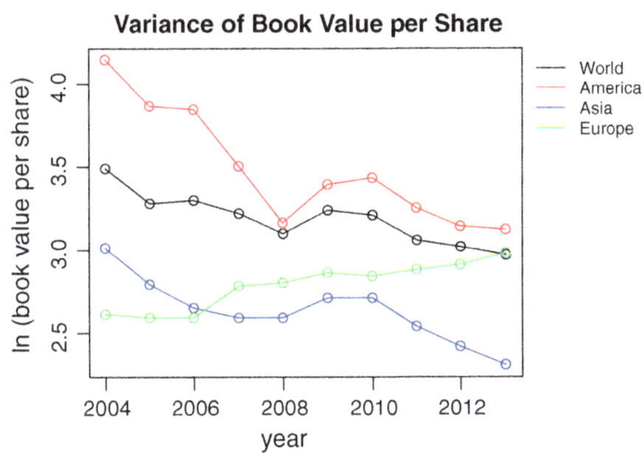

Fig. 6.14 Changes in variance of book value per share

6.6 Concluding Remarks

In this study, we investigated whether the distribution of share price follows a power law distribution at the regional level. We found that the distribution of share price follows a power law distribution for all regions and that the regional power law exponents are quite diverse over the range 0.98–3.42.

Using a panel regression model for share price in which share price is the dependent variable and dividends per share, cash flow per share, and book value per share are the explanatory variables, we estimated the theoretical value. A two-way fixed effects model was identified as the best model for all regions. Since the two-way fixed effects model includes an individual fixed effects term and a time fixed effects term in its error term, we were able to produce a measure of company fundamentals by removing the time fixed effect from the theoretical value. The fact that the two-way fixed effects model was selected for all regions allowed us to consistently compute the fundamentals at the regional level in the same way.

The model was found to have quite a high power to explain share price at the regional level, showing large R^2 values in the range of 0.95 to 0.98. As a result, we were able to show that the distributions of theoretical value coincide with the distributions of share price for all regions.

Investigating the distribution of company fundamentals at the regional level, we found that the distribution follows a power law distribution for all regions. In addition, the distribution of fundamentals was found to coincide with the distribution of share price for most regions.

Furthermore, the distributions of the financial indicators per share that were used as explanatory variables in the econometric model were shown to follow a power law distribution. From these results, it can be said that fundamentals consisting of the financial indicators per share and representing corporate value are the essential determinants of share price.

We found that the power law exponents were quite diverse by region. However, the power law exponents for Europe were close to those for the world. We surmised that the power law distribution for the world was heavily influenced by the European data. The power law exponents were not extensively examined at the regional level in this study but will be the theme for future studies.

Acknowledgements This research was supported by JSPS KAKENHI Grant Number 2538404, 2628089.

References

Anirban Chakraborti, Ioane Muni Toke, Marco Patriarca & Frédéric Abergel (2011a) Econophysics review: I. Empirical facts, Quantitative Finance, Volume 11, Issue 7 991–1012.

Anirban Chakraborti, Ioane Muni Toke, Marco Patriarca & Frédéric Abergel (2011b) Econophysics review: II. Agent-based models, Quantitative Finance, Volume 11, Issue 7 1013–1041.

Axtell R L. (2001) Zipf's Distribution of U.S.Firm Sizes, Science, Vol. 293, 1818–1820.
Clauset A., Shalizi C.R., and Newman M. E. (2009) Power law distributions in empirical data, SIAM Review, Vol. 51, No.4, pp.661–703.
Čížek P. Härdlle, W. Weron, R. (2005) Statiscal tools for finance and insurance. Springer, Berlin.
D'Agostino, R. B. and Stephans, M.A. (1986) Goodness-of-Fit Techniques, Marcel Dekker New York.
Gabaix, X. (1999). Zipf's Law for Cities: An Explanation. The Quarterly Journal of Economics, 739–767.
Kaizoji T. and Miyano M. (2016a) Why does power law for stock price hold? Journal of Chaos Soliton and Fractals. (press).
Kaizoji T. and Miyano M. (2016b) Stock Market Crash of 2008: an empirical study on the deviation of share prices from company fundamentals, mimeo, to be submitted.
Kaizoji T. and Miyano M. (2016c) Zipf's law for company fundamentals and share price, mimeo, to be submitted.
Newman M. E.J. (2005) Power laws, Pareto distributions an Zipf's law, Contemporary Physics, Volume 46, Issue 5, 323–351.
Woodlridge J. M. (2010) Econometric Analysis of Cross Section and Panel Data, MIT press.

Chapter 7
Record Statistics of Equities and Market Indices

M.S. Santhanam and Aanjaneya Kumar

Abstract Record events in a time series denotes those events whose magnitude is the largest or smallest amongst all the events until any time N. Record statistics is emerging as another statistical tool to understand and characterise properties of time series. The study of records in uncorrelated time series dates back to 60 years while that for correlated time series is beginning to receive research attention now. Most of these investigations are aided by the applications in finance and climate related studies, primarily due to relatively easy availability of long measured time series data. Record statistics in respect of empirical financial time series data has begun to attract attention recently. In this work, we first review some of the results related to record statistics of random walks and its application to stock market data. Finally, we also show through the analysis of empirical data that for the market indices too the distribution of intervals between record events follow a power law with exponent lying the range 1.5–2.0.

7.1 Introduction

Record events have a popular appeal and generally enjoy continuous media attention. Record breaking events such as the highest or lowest temperature ever reached in a city, largest magnitude of rainfall ever recorded, accidents with biggest number of causalities, highest opening weekend collection of movies, biggest fall or rise in stock market indices, unparalleled sport performances always have curiosity value as seen by the popularity of Guiness book of world records (Guinness World Records 2016). For example, India's fourth largest city Chennai received nearly 1049 mm of rainfall during November 2015 almost breaking a century old record (Wikipedia entry 2015)

M.S. Santhanam (✉) · A. Kumar
Indian Institute of Science Education and Research,
Dr. Homi Bhabha Road, Pune 411008, India
e-mail: santh@iiserpune.ac.in

A. Kumar
e-mail: kumar.aanjaneya@students.iiserpune.ac.in

F. Abergel et al. (eds.), *Econophysics and Sociophysics: Recent Progress and Future Directions*, New Economic Windows,
DOI 10.1007/978-3-319-47705-3_7

leading to closure of its busy airport and consequent disruption of economic activity estimated at billions of dollars. This episode also included the record-breaking 24-h rainfall in chennai's over 200-year old history of meteorological observations. In view of the massive economic impact of such large record events, it is important to understand their statistical properties and, if possible, predict them. In general, the idea of record events has proven useful in other domains of physics too. In physics literature, records statistics is emerging as an important area of research especially in the context of complex systems and has found applications, for instance, in understanding magnetization in superconductors and as an indicator of quantum chaotic effects (Oliveira et al. 2013). In general, understanding the statistics of record events in complex systems would lead to a better characterization and appreciation of the extremal properties of real systems. In this work, we present empirical results for the record statistics of stock market data, both for equities and indices.

7.2 Record Statistics

Let $x_t, t = 1, 2, 3, \ldots T$ denote a discretely sampled stochastic and univariate time series. The record events are those that are either larger than all the previous values or smaller than the previous values. An event at $t = \tau$ would constitute an upper record event if $x_\tau > \max(x_1, x_2, \ldots, x_{\tau-1})$. Similarly, in the case of lower record event at time $t = \tau$, $x_\tau < \min(x_1, x_2, \ldots, x_{\tau-1})$. In Fig. 7.1a, we display the daily closing values of NYSE AMEX composite index (XAX) for the years from 1996 to 2015. In

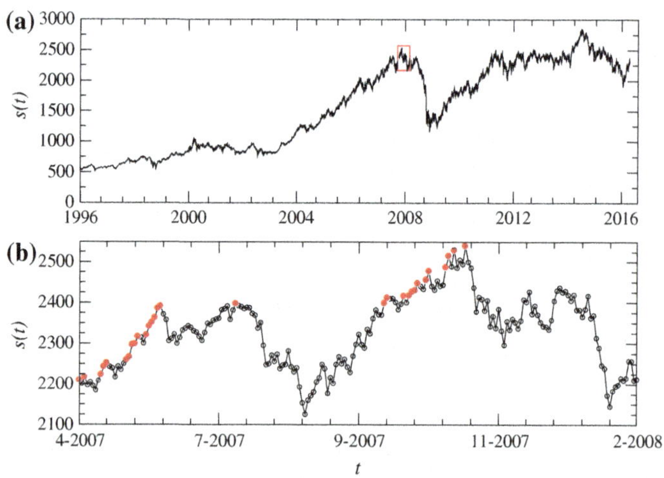

Fig. 7.1 **a** Daily closing values of the NYSE AMEX composite index (XAX) shown for the years from 1996 to 2015. **b** A part of the figure **a** indicated by the *red box* is enlarged. The *red (filled) circles* denote the occurrence of upper records

Fig. 7.1b, we have marked the position of the record events for a short portion of the data. The time interval between two successive occurrences of record events is called the record age. The principal results in this paper relate to the statistical properties and the distribution of record age.

If the time series x_t is uncorrelated, many properties of record events are already known. For instance, the distribution of record age for an uncorrelated time series of length N can be shown to be (Schmittmann and Zia 1999)

$$f_N(m) = 1/m \tag{7.1}$$

and remarkably this is independent of N and the distribution $P(x)$ of underlying time series. Physical content of Eq. 7.1 implies that the probability of a record event taking place at any instant does not depend on how much of data we have already measured for that process.

However, most of real-life observations pertaining to data from stock markets, geophysical processes (such as temperature, earthquake etc.) and physiological processes (such as heart beat intervals, electroencephalagram measurements etc.) are correlated (Doukhan et al. 2003). Often, such variables display $1/f$ noise type power spectrum, a signature of complexity in the system (Turcotte and John Rundle 2002). For instance, a large body of results show that the heart beat intervals display $1/f$ noise (Plamen Ch. Ivanov 2001). It is then natural to ask as to what happens to record statistics for such correlated systems? This is currently one of the important questions pursued in the record statistics literature.

7.3 Record Statistics of Random Walks

In physics, random walk is a fundamental model that forms the basis for our understanding of diffusion processes in general (Rudnick and Gaspari 2004). According to this model, one can imagine a walker choosing next position to hop to through probabilistic means. This has been widely applied in many areas including in finance (Fama 1995; Ruppert 2006). A discrete time version of random walk problem is given by

$$y_{i+1} = y_i + \xi_i, \qquad i = 0, 1, 2, \ldots \tag{7.2}$$

in which i represents discrete time and ξ_i is a random number from a suitable distribution $\phi(\xi)$. In this, the positions y_i of the random walker are correlated and hence this model being reasonably simple lends itself for analysis from record statistics point of view.

In the last few years, the record statistics of random walker was studied analytically (Majumdar and Ziff 2008; Majumdar 2013). The principal result is that for N-step random walk, if the distribution $\phi(\xi)$ is both symmetric and continuous, the mean number of records and the mean record age for large N are, respectively,

$$\langle m \rangle \propto \sqrt{N}, \qquad \text{and} \qquad \langle r \rangle \propto \sqrt{N}. \tag{7.3}$$

This result is based on the use of Sparre-Andersen theorem and is independent of the form of $\phi(\xi)$ except for the requirement of independence and continuity (Majumdar and Ziff 2008; Majumdar 2013). However, unbiased random walk in Eq. 7.2 is not suitable for stock market data applications since most stock data generally have a net drift. In order to account for this, earlier results in Majumdar and Ziff (2008) were extended to describe the model $y_{i+1} = y_i + \xi_i + c$, where ξ is a random variable from a Gaussian distribution $G(0, \sigma)$ and c is the constant drift. In this case, in Wergen et al. (2011), the mean number of records was obtained as

$$\langle m_N \rangle \approx \frac{2\sqrt{N}}{\sqrt{\pi}} + \frac{\sqrt{2}c}{\pi\sigma}\left(N \arctan(\sqrt{N}) - \sqrt{N}\right), \tag{7.4}$$

provided $c/\sigma \ll 1/\sqrt{N}$. This result was compared with the S&P 500 index data during the years 1990–2009. As might be expected, the unbiased random walk does not correctly capture the mean number of records (see Fig. 7.2). The result of biased model in Eq. 7.4 provides a substantial improvement over that of unbiased random walk model as seen in the somewhat close agreement (shown in Fig. 7.2) between the empirical data and the analytical result in Eq. 7.4. Though there is scope for improvement, it appears reasonable to state that random walk model with a drift provides a better description of mean number of records than the unbiased random walk model. Exact or asymptotic results for the mean record number, mean record

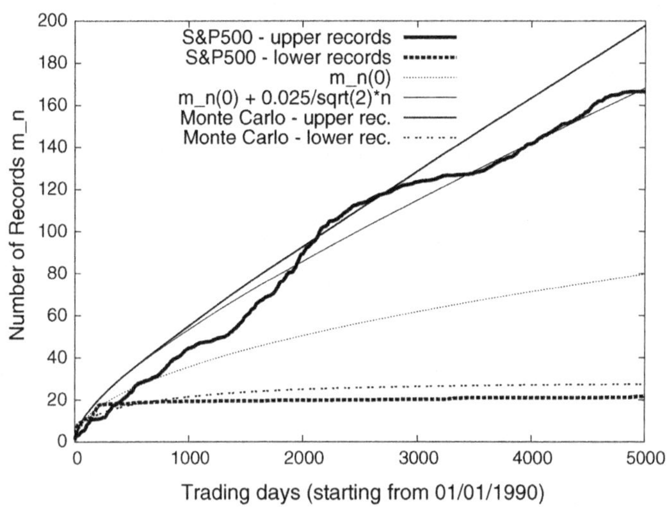

Fig. 7.2 Mean number of record events for the stocks constituting the S&P index for the years 1990 to 2009 compared with the model of random walk with drift (Wergen et al. 2011). (Reproduced with permission from American Physical Society)

age, shortest and longest record ages have been obtained for the biased random walk model in Majumdar (2013). A more systematic comparison of the results from random walk and linear time series models with the S &P 500 stocks and index data reveals important information about the deviation between the two (Wergen 2014). In this work, autoregressive GARCH (1,1) model was found to reasonably model the record breaking daily prices of stocks in addition to obtaining the distribution of number of records up to some time N.

In the last few years, more variants of random walk problem have been studied. They include, record statistics with many random walkers (Wergen et al. 2012), random walks with measurement error (Edery et al. 2013) and continuous time random walks (Sabhapandit 2011). We also point out record statistics is being investigated in the context of climate change and to understand its effect on record temperature occurrences (Redner and Petersen 2006; Newman et al. 2010). For such geophysical phenomena, random walk is not the suitable framework. In Redner and Petersen (2006), analytical estimates for mean number of records and mean time between successive records is obtained by making statistical assumptions about the distribution of temperature values. These estimates differ considerably from the random walk results but are seen to be reasonably valid for measured temperature data (Redner and Petersen 2006; Newman et al. 2010).

7.4 Record Age Distribution for Stock Data

The typical record age can be written as, $\langle r\rangle \sim N/\langle M\rangle$. For the unbiased random walk, we get $\langle r\rangle \sim N/\sqrt{N} = \sqrt{N}$. In this work, we are interested in the distribution of record ages. Analytical results for the distribution of record age is as yet not known. We briefly review our recent largely numerical work on this problem (Sabir and Santhanam 2014). The main result obtained from the analysis of time series of stock data is that the record age distribution is consistent with power-law of the form (Sabir and Santhanam 2014),

$$P(r) \sim A\, r^{-\alpha} \tag{7.5}$$

where the exponent $1.5 \leq \alpha \leq 1.8$. Further, A is the normalization constant that can be expressed as a harmonic number $H_{N,\alpha}$.

For a time series of length N, unity is the lower bound on record age and any record age longer than the length of the time series cannot be resolved. As shown in Fig. 7.3a, the record ages for IBM stock displays values spanning about two orders of magnitude. Its distribution $P(r)$ shown in Fig. 7.3b, for IBM, HPQ and XOM stocks, displays power-law form with the exponent $\alpha \sim -1.623 \pm 0.081$. The analysis of 19 stocks reveals that the exponent of the record age distribution α satisfies $1.5 \leq \alpha \leq 1.8$. This is shown in Fig. 7.4 with the exponent α obtained as maximum likelihood estimate from empirical data. We emphasise that the distribution $P(r)$ in Eq. 7.5 is independent of the length N of data. Unlike the quantities like the mean number of

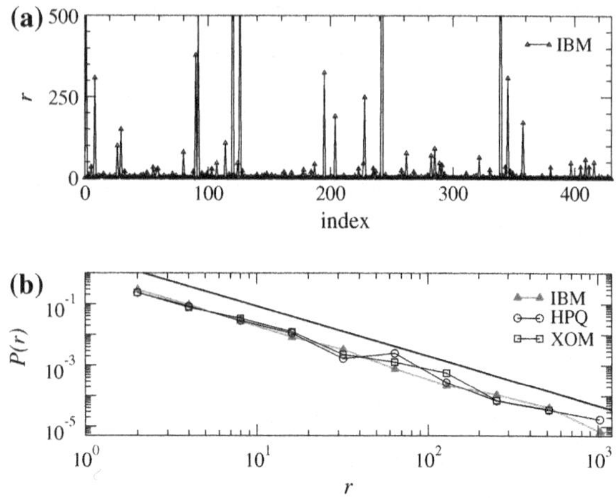

Fig. 7.3 **a** Record ages (in days) calculated from IBM stock data. **b** The distribution of record ages for three stocks. The best fit *solid line* in (**b**) has slope -1.58 ± 0.15 (Sabir and Santhanam 2014). (Reproduced with permission from American Physical Society)

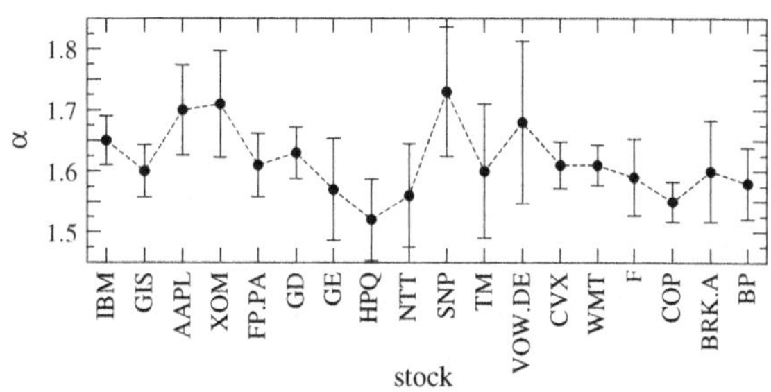

Fig. 7.4 Exponent α obtained as maximum likelihood estimate from empirical stock data (Sabir and Santhanam 2014). (Reproduced with permission from American Physical Society)

records and typical record age which depend on the length of data (see Eqs. 7.3–7.4), we can regard record age distribution as characterising the record events in a system independent of the data length.

As a framework to understand these results, the geometric random walk (GRW) model is considered, which is suitable as a model for the time evolution of stocks. The GRW model is given by $y_{i+1} = y_i \ \exp(\xi_i)$, in which ξ is Gaussian distributed $G(\mu, \sigma)$ with mean μ and standard deviation σ. In the context of stock market modelling, one of the virtues of GRW model is that the variable $y_i \geq 0$ as we expect the stock prices to be positive semi-definite. Secondly, the log-returns $\log(y_{i+1}/y_i)$ are normally distributed, a feature seen in many empirical stock returns over a wide range of time scales (Ruppert 2012). In Sabir and Santhanam (2014), it was shown that GRW is suitable to model the record statistics of real stock data. Indeed, the exponent of the record age distribution of GRW time series is seen to be $\approx$1.61. We also note that the distribution of interval between two successive record events for the case of temperature data has been found to be a power law as well (Redner and Petersen 2006).

7.5 Record Age Distribution for Market Indices

Index of a stock market is generally calculated as the weighted average market capitalization of the stocks that make up the index. It serves as a single number indicator of the evolution of equities in a market. Market surges and crashes are indicated by the dynamics of market index and in this sense it reflects the overall direction of the market. In August 2011, S &P 500 index lost nearly 7 % and more recently Chinese market suffered downfall in 2015–16. Statistical properties of such record events should be of interest in the context of recurring market surges and crashes. This motivates the study of record events in the stock market indices.

In the rest of the paper, we use publicly available data of major indices from http://www.finance.yahoo.com, the details of which are provided in the Appendix. In Fig. 7.5, we show the record age distribution for three different market indices, namely, Nasdaq composite index, FTSE and Nikkei. In all the three cases the record age distribution is a power-law well described by $P(r) \propto r^{-\alpha}$. The exponent values, respectively, are 1.72 ± 0.088, 1.66 ± 0.128 and 1.78 ± 0.185. Similar to the case of stock data, this result is independent of the length of data considered for analysis. However, mean record age can be dependent on the length N of data. Given that $P(r) \propto r^{-\alpha}$, it is straightforward to see that $\langle r \rangle \propto N^{2-\alpha}$. If $\alpha \sim 1.5$, then $\langle r \rangle \propto \sqrt{r}$, a result indicated by random walk based analysis (Majumdar and Ziff 2008). In our empirical analysis, the value of the exponent lies in the range $1.5 \leq \alpha \leq 2$ as displayed in Fig. 7.6 for a collection of market indices. This is similar to the result presented in Fig. 7.4 for equities. In practice, this implies that both for the indices and stocks, the mean waiting time for the next record event depends on N except if the exponent is $\alpha = 2$.

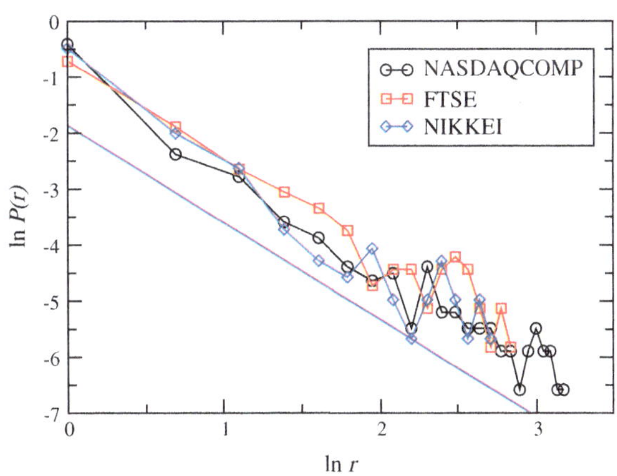

Fig. 7.5 Record age distribution for three market indices. The *solid line* is mean of the best fit lines with slope −1.72 and is shown as a guide to the eye

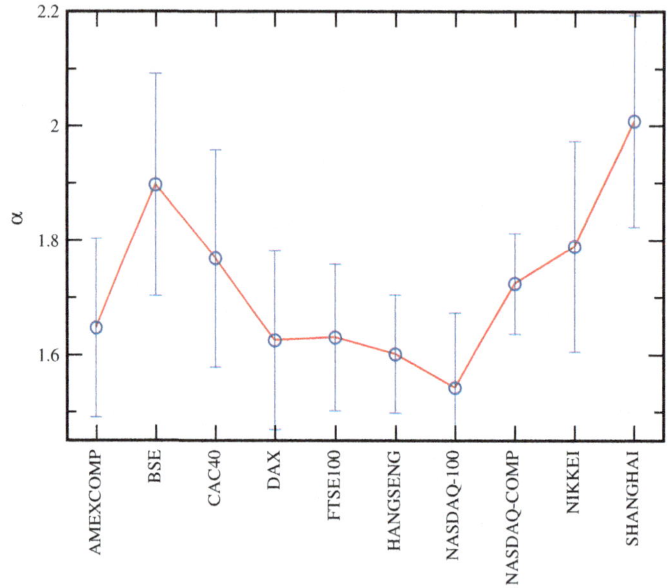

Fig. 7.6 Exponent α displayed for various market indices shown on x−axis. The error bars represent the uncertainties computed by the linear regression

7.6 Summary and Conclusions

In summary, we have reviewed some recent results related to the record statistics of random walks and biased random walks and in particular their application to the record events in the time series data of equities and market indices. In this context, main questions that have received attention are the mean number of records until time N and the mean record age. For stock market applications, mean record age indicates the average waiting time for the next record event to take place.

In this work, going beyond just the mean record age, we study the record age distribution. Earlier results had shown that the record age distribution for the record events in equities is a power of the form $P(r) \sim Ar^{-\alpha}$, where $1.5 \leq \alpha \leq 1.8$. In this work, we analyse the data of indices from ten major markets for the record events and show that record age distribution is a power-law with exponent in nearly the same range as the case for equities. A significant aspect of this result record age distribution does not depend on the length N of data being considered. This is in constrast with the mean number of records and mean record age, which explicitly depend on N. It would be interesting to provide further analytical insight into these problems.

Acknowledgements AK would like to thank DST-INSPIRE for the fellowship. We acknowledge the useful data provided from http://finance.yahoo.com without which this work would not have been possible.

Appendix

The details of the indices data used are given here. The data is available in the public domain and can be accessed from http://finance.yahoo.com.

Index	Length of data	Years covered
AMEX Composite	5117	1996–2016
BSE	4650	1997–2016
CAC40	6628	1990–2016
DAX	6434	1990–2016
FTSE100	8412	1984–2016
HANGSENG	7291	1987–2016
NASDAQ 100	7706	1985–2016
NASDAQ composite	11404	1971–2016
NIKKEI	7958	1984–2016
SHANGHAI	6465	1990–2016

References

P. Doukhan, G. Oppenheim and M. S. Taqqu, *Theory and Applications of Long-Range Dependence*, (Springer, 2003).

Yaniv Edery, Alexander B. Kostinski, Satya N. Majumdar, and Brian Berkowitz, Phys. Rev. Lett. **110**, 180602 (2013).

Eugene F. Fama, Fin. Anal. J. **51**, 75 (1995); see also, Andrew W. Lo and A. Craig MacKinlay, Rev. Financ. Stud. **1**, 41 (1988);

Guinness World Records, http://www.guinnessworldrecords.com. Cited 30 Apr 2016.

Plamen Ch. Ivanov *et. al.*, Nature **399**, 461 (1999); Plamen Ch. Ivanov *et. al.*, Chaos **11**, 641 (2001).

Satya N. Majumdar and Robert M. Ziff, Phys. Rev. Lett. **101**, 050601 (2008).

Satya N. Majumdar, Grégory Schehr and Gregor Wergen, J. Phys. A **45**, 355002 (2012); Alan J Bray, Satya N Majumdar, Grégory Schehr, Adv. Phys. **62**, 225 (2013).

W. I. Newman, B. D. Malamud and D. L. Turcotte, Phys. Rev. E **82**, 066111 (2010); G. Wergen and J. Krug, EPL **92**, 30008 (2010); S. Rahmstorf and D. Coumou, PNAS **108**, 17905 (2011); Gregor Wergen, Andreas Hense, Joachim Krug, Clim. Dyn. **22**, 1 (2015); M. Bador, L. Terray and J. Boe, Geophys. Res. Lett. **43** 1 (2016).

L. P. Oliveira, H. J. Jensen, M. Nicodemi and P. Sibani, Phys. Rev. B **71**, 104526 (2005); P. Sibani, G. F. Rodriguez, and G. G. Kenning, Phys. Rev. B **74**, 224407 (2006); Shashi C. L. Srivastava, A. Lakshminarayan and Sudhir R. Jain, Europhys. Lett. **101**, 10003 (2013).

S. Redner and Mark R. Petersen, Phys. Rev. E **74**, 061114 (2006).

Joseph Rudnick and George Gaspari, *Elements of the Random Walk*, (Cambridge University Press, 2004).

D. Ruppert, *Statistics and Finance: An Introduction*, (Springer, New York, 2006).

David Ruppert, *Statistics and Data Analysis for Financial Engineering*, (Springer, New York, 2011); Frank J. Fabozzi, *Encyclopedia of Financial Models*, 1st ed. (Wiley, New York, 2012).

Sanjib Sabhapandit, EPL **94**, 20003 (2011).

Behlool Sabir and M. S. Santhanam, Phys. Rev. E **90**, 032126 (2014).

B. Schmittmann and R. K. P. Zia, Am. J. Phys. **67** 1269 (1999); Joachim Krug, J. Stat. Mech. (Theory and Expt) P07001 (2007); Gregor Wergen, J. Phys. A : Math. Theor. **46**, 223001 (2013).
D. L. Turcotte and John Rundle, PNAS **99**, 2463–2465 (2002) and all other papers in this issue of PNAS.
Wikipedia entry on Chennai floods of 2015. https://en.wikipedia.org/wiki/2015_South_Indian_floods
Gregor Wergen, Miro Bogner and Joachim Krug, Phys. Rev. E **83**, 051109 (2011).
Gregor Wergen, Satya N. Majumdar, and Grégory Schehr, Phys. Rev. E **86**, 011119 (2012).
Gregor Wergen, Physica A **396**, 114 (2014).

Chapter 8
Information Asymmetry and the Performance of Agents Competing for Limited Resources

Appilineni Kushal, V. Sasidevan and Sitabhra Sinha

Abstract While mainstream economic theory has been primarily concerned with the behavior of agents having complete information and perfect rationality, it is unlikely that either of these assumptions are valid in reality. This has led to the development of theories that incorporate bounded rationality and also to the study of the role of information in economic interactions (information economics). In particular, information asymmetry, where all the agents do not have access to the same information has aroused much attention, as it has potential to significantly distort economic outcomes resulting in the failure of the market mechanism. It is often assumed that having more data than others gives agents a relative advantage in their interactions. In this paper we consider the situation where agents differ in terms of the granularity (as well as the quantity) of the information that they can access. We investigate this in the framework of a model system comprising agents with bounded rationality competing for limited resources, viz., the minority game. We show that there is no simple relation between the amount of information available to an agent and its success as measured by payoffs received by it. In particular, an agent having access to a much coarser-grained information (that is also quantitatively less) than the rest of the population can have a relative advantage under certain conditions. Our work shows that the success of individual agents can depend crucially on the relative fraction of the population that uses information of a specific type.

A. Kushal (✉)
Indian Institute of Science, C V Raman Road, Bangalore 560012, India
e-mail: akushalstar@gmail.com

V. Sasidevan · S. Sinha
The Institute of Mathematical Sciences, CIT Campus, Chennai 600113, India
e-mail: sasidevan@imsc.res.in

S. Sinha
e-mail: sitabhra@imsc.res.in

F. Abergel et al. (eds.), *Econophysics and Sociophysics: Recent Progress and Future Directions*, New Economic Windows,
DOI 10.1007/978-3-319-47705-3_8

8.1 Introduction

Economic phenomena provide some of the most fascinating examples of non-trivial collective behavior emerging from the interactions between components of a complex adaptive system (Sinha et al. 2010). In particular, markets provide an ecosystem for a large number of individuals to exchange goods and/or services for satisfying their mutual needs in a self-organized manner (Miller and Page 2007). The fact that markets in the real world operate fairly efficiently most of the time despite being subject to noise, external shocks and lack of complete information on the part of all agents, has led to many attempts at explaining the underlying mechanisms. Unlike mainstream economic models that assume complete information and rationality for the agents, in reality there is often significant asymmetry among the market participants in terms of information available to them, as well as, their capacity to analyze this information (Simon 1955). Information asymmetry, in particular, can lead to significant distortions in interactions between buyers and sellers, and may result in failure of the market mechanism (Stiglitz 2000).

Agents can—and often do—have access to information which differ not only in quantitative terms ("how much data?") but also qualitatively ("what type of data?"). For instance, in a financial market, agents can speculate on the price variation of individual stocks or invest in funds whose values are linked to fluctuations in the market indices, the latter representing information that is much more coarse-grained than the former. We can ask whether having more detailed information like the former necessarily translate into an advantage for the corresponding agents when they are pitted against agents using the latter type of information having a lower resolution. In general, we can have information at differing levels of granularity which the agents make use of in their strategies for attaining specific economic goals.

In this paper, we use a formal simplified model of a system in which heterogeneous agents with bounded rationality compete for limited resources and who have access to either of two very distinct types of information about the outcomes of their previous interactions. These different information, which represent the two extreme cases of granularity possible for the collective state of the model, are used by the corresponding type of agents for the same purpose, viz., predicting the future outcome. We show that there is no simple relation between the amount of information available to an agent and its success as measured by payoffs received by it. In particular, an agent having access to coarse-grained information (that is also quantitatively less) than the rest of the population can have a relative advantage under certain conditions. In general, which type of agent will fare better depends upon the exact composition of the population, as well as, the amount of information available to agents of each type. Our work shows that the success of individual agents can depend crucially on the relative fraction of the population that uses information of a specific type. This is a novel systems-level phenomenon that emerges essentially in a self-organized manner from interactions between a large number of heterogeneous entities.

8.2 The Model

Our intention is to investigate the result when agents having access to qualitatively different information interact with each other in a complex adaptive system. For this purpose, we shall consider a setting where agents choose actions based on strategies that use information about past outcomes. One of the simplest models implementing this paradigm is the Minority Game (MG) (Challet and Zhang 1997) inspired by the El Farol Bar problem proposed by Arthur (1994). In this game, an odd number N of agents have to choose between two options (A and B, say) at each round, independently and simultaneously, with those on the minority side winning—corresponding to a payoff 1, say—and those on the majority side losing—corresponding to a payoff 0 (for a concise account of MG see, e.g., Moro 2004). We augment the basic MG model by having two different types of agents distinguished by the type of information that they use.

The first type of agent is identical to that in the conventional Challet-Zhang Minority Game (CZMG) (Challet and Zhang 1997). These agents make their selection based on the common information about the identity of the side (whether A or B) that was occupied by the minority group on each of the previous $m1$ rounds. The information is thus a binary string of length $m1$. A strategy used by an agent is a rule that informs an agent whether to select A or B the next day for all possible past contingencies. The total number of such strategies that are possible is thus $2^{2^{m1}}$.

The second type of agent we consider uses detailed (as opposed to binary) information about past outcomes (Sasidevan 2016) (also see Dhar et al. 2011; Sasidevan and Dhar 2014). Such Detailed Information Minority Game (DIMG) agents have access to information about the exact number of agents who opted for a particular choice in the previous $m2$ rounds (Fig. 8.1). The information is therefore a string of length $m2$, where each entry is an integer between 0 and N. Like the CZMG agents, they have strategies that use this detailed information to make predictions about the choice that will be made by the minority in each round. The total number of such strategies possible is $2^{(N+1)^{m2}}$.

Once we have decided on the composition of the population, i.e., the relative numbers of CZMG and DIMG agents, the game evolves according to the following rules. Each agent initially chooses a small number (typically 2, as in this paper) of strategies at random from the set of all possible strategies corresponding to their type (CZMG or DIMG). An agent can measure the performance of each of its strategies by assigning a score based on how well they predicted the option chosen by the minority in the past. At each round, an agent uses the strategy with the highest score that is available to it.

It is known that in the conventional MG, agents having large enough memory size self-organize into a state where the fluctuation in the number choosing a particular option about the mean value ($N/2$ due to the symmetry between A and B) is minimized. This results in the population as a whole doing better (i.e., it is globally efficient) than the case in which agents choose randomly between A and B with equal probability. Global efficiency is maximized for a critical value of memory size

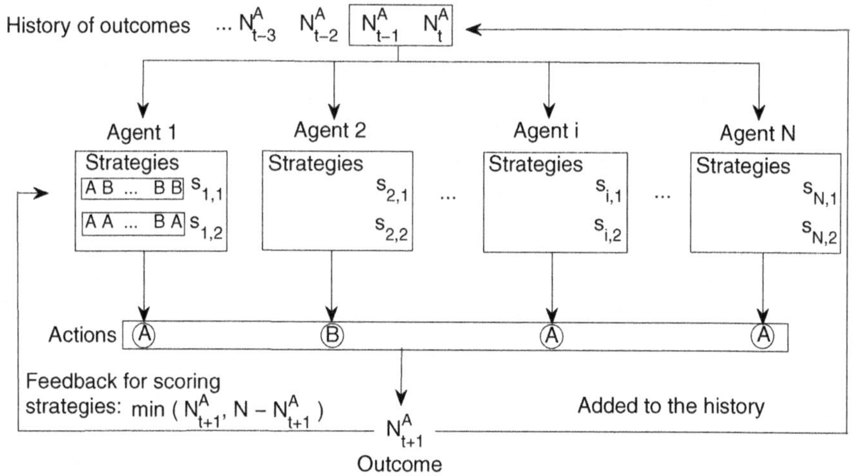

Fig. 8.1 A schematic representation of the Minority Game involving agents having access to detailed information (DIMG), i.e., they use the exact number of agents who opted for a particular choice (say A), to predict the option that will be chosen by the minority at each iteration. In the figure it is assumed that agents have a memory size of 2, i.e., they have detailed information from the two preceding rounds, viz., N_t^A and N_{t-1}^A. Each of the N agents have two possible strategies $s_{i,1}$ and $s_{i,2}$ at hand and use the one with the best score in any given round. After each round, their aggregate action N_{t+1}^A is added to the history of outcomes and they update their strategy scores based on the identity of the winning choice (i.e., the option chosen by the minority)

of agents, $m_c \sim \log_2 N$ (Challet and Zhang 1998; Challet et al. 2000, 2005). When the memory size is $\ll m_c$, the agents exhibit herding behavior where most of them choose the same option in a given round, resulting in very large fluctuations about the mean. For such a situation, the individual payoffs are extremely low and the system is also globally inefficient—even compared to simple random choice behavior.

8.3 Results

We now look at the results of interactions between agents having access to qualitatively different information, viz., detailed, i.e., exact number opting for a particular choice, versus binary, i.e., the identity of the winning choice. We first focus on the simplest cases where a single agent having access to binary information interacts with a population of agents having detailed information. We then look at the opposite case where an agent using detailed information is pitted against a population of agents using only binary information. The quality of information available to these two types of agents represent the two extremes of data granularity. The agents use this different types of information for the same purpose, i.e., predict the outcome of the game, viz., the option chosen by the minority, at each iteration. Finally, we also look at the general case where the fraction of agents having access to binary or detailed information is varied over the entire range between 0 and 1.

8.3.1 Single DIMG Agent Interacting with $N - 1$ CZMG Agents

We introduce a DIMG agent, i.e., one who has access to information about the exact number of agents who opted for choice A over the previous $m2$ iterations, in a population where the remaining $N - 1$ agents only know of the identity of the winning choice over the preceding $m1$ iterations, i.e., they are CZMG type agents with memory length $m1$. Both types of agents use the information available to them to predict the option that will be chosen by the minority in each round. The resulting performance of the agents (measured in terms of their average payoff) is shown as a function of the memory length of the CZMG agents in Fig. 8.2. The behavior is seen to be almost independent of the memory length of the DIMG agents [compare panels (a) and (b) of Fig. 8.2]. Furthermore, increasing the size of the population N does not change the qualitative nature of the payoffs as a function of $m1$ beyond shifting the extrema towards higher values of $m1$.

The single agent having access to detailed information clearly has an advantage over the other type of players when the memory length $m1$ of the latter is small ($m1 < \log_2 N$). The payoff of the DIMG agent decreases with increasing $m1$ and is lowest when the emergent coordination among the CZMG agents is highest—which, according to the well-known result of MG, occurs when $2^{m1} \simeq 0.337N$ (Challet et al. 2000). The superior performance of the lone agent with detailed information when

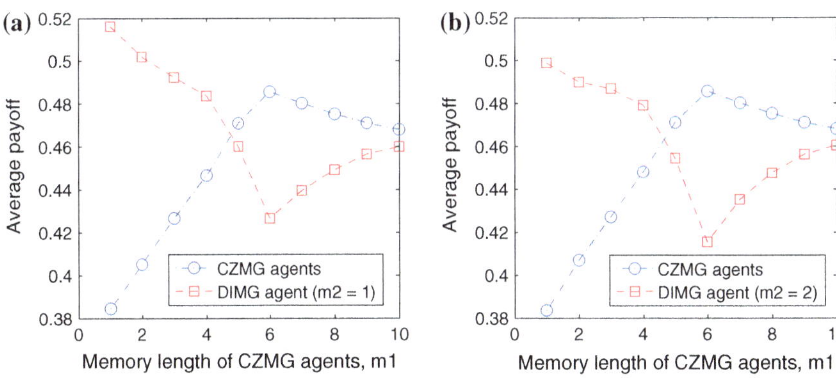

Fig. 8.2 Average payoffs of different types of agents as a function of the memory length $m1$ of agents having binary information (CZMG) when $N - 1$ such agents interact with a single agent having detailed information (DIMG). Payoffs are averaged over 10^4 iterations in the steady state and over 250 different realizations with $N = 127$. The variation of the payoffs with $m1$ shows a similar profile for different memory lengths **a** $m2 = 1$ and **b** $m2 = 2$ of the DIMG agent

$m1 \ll \log_2 N$ can be understood as related to the predictable collective behavior of the CZMG agents. As is well known, in this regime, the number of CZMG agents choosing a particular option (and the resulting minority choice) changes in a periodic manner, fluctuating between very low and high values (Challet and Zhang 1998). As a result, the CZMG agents receive relatively lower payoffs. It is perhaps not surprising that an agent who does not move in synchrony with the large number of CZMG agents will do better. This is also true for an agent who chooses between the options at random. We note that the DIMG agent performs somewhat better than random (figure not shown), presumably because of the adaptive way in which it chooses between the options.

As $m1$ increases, the collective behavior of the CZMG agents loses its predictability resulting in a gradual decrease in the payoff of the single DIMG agent. Beyond the minimum around $m1 \sim \log_2(0.337N)$, we observe that the payoff of the DIMG agent again starts increasing and approaches the decreasing payoff of the CZMG agents, eventually converging to the payoff expected for random choice for high enough $m1$. The behavior of CZMG agents as a function of $m1$ is well-understood theoretically (Challet et al. 2000; Hart et al. 2001; Coolen 2004). We note that the trend of the payoff of the DIMG agent as a function of $m1$ mirrors that of the CZMG agents.

8.3.2 Single CZMG Agent Interacting with $N - 1$ DIMG Agents

One may naively argue that the relative advantage of the DIMG agent in making predictions when they interact with CZMG agents having small memory size $m1$ (as described above) may be understood as resulting from the former having quantitatively more information (e.g., measured in terms of bits) available at their disposal. However, such an argument will fail to explain the behavior seen in the other extreme case where a single agent having access to only binary information (i.e., the minority choice) over the preceding $m1$ iterations is introduced in a population where the remaining agents have information about the exact number of agents opting for a particular choice over the preceding $m2$ iterations. The resulting performance of the agents (measured in terms of their average payoff) is shown as a function of the memory length of the CZMG agent in Fig. 8.3 for two different population sizes. Unlike the other extreme case, we note that the results seem to depend on the memory length $m2$ of the DIMG agents (compare between panels (a, c) and (b, d) of Fig. 8.3).

The most striking feature for $m2 = 1$ is that the single CZMG agent with $m1 > 1$ performs better than the rest of the population for small $m1$, i.e., in a regime where it actually has quantitatively much less information than the other agents. This is true for different population sizes (shown for $N = 127$ and 255 in Fig. 8.3) although the range of $m1$ for which the CZMG agent has an advantage over the DIMG agents does depend upon N. The implication of this result is far reaching as it suggests that

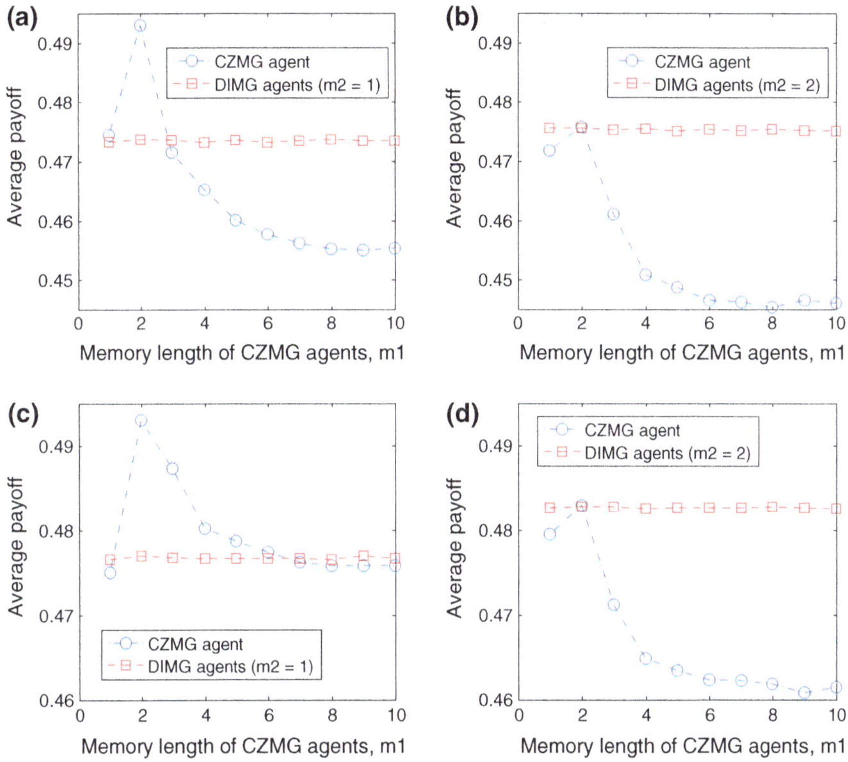

Fig. 8.3 Average payoffs of different types of agents as a function of the memory length $m1$ of an agent having binary information (CZMG) when such an agent interacts with $N - 1$ agents having detailed information (DIMG). Payoffs are averaged over 10^4 iterations in the steady state and over 250 different realizations with **a**–**b** $N = 127$ and **c**–**d** $N = 255$. The variation of the payoffs with $m1$ shows different profiles for different memory lengths **a**, **c** $m2 = 1$ and **b**, **d** $m2 = 2$ of the DIMG agents

just having quantitatively more information does not necessarily translate into better performance in predicting the future outcomes. Instead, the success of an agent in an ecosystem of agents using different types of information depends on being able to "stand apart from the crowd" even when that means using less amount of data than the others, allowing it to take advantage of predictable patterns in the collective behavior of the rest of the agents. Thus, striving to collect and process ever increasing quantities of data in the hope of making more accurate predictions in complex adaptive systems such as financial markets may actually be counter-productive.

We now consider the case where the memory length of the DIMG agents is increased to $m2 = 2$ (Fig. 8.3b, d). We observe that in this case the CZMG agent has no advantage over the rest of the population regardless of its memory length $m1$. Just as the CZMG agents achieve maximum emergent coordination when their memory length is of the order of $\log_2(N)$, it is known that the DIMG agents achieve the same for $m2 = 2$ independent of N (Sasidevan 2016). Thus, using our arguments in the other extreme case, we expect that the single CZMG agent will not have any advantage over the optimally coordinated DIMG agents. If $m2 > 2$, the behavior of a group of DIMG agents is indistinguishable from agents who randomly choose between the options. Note that there is the possibility that if more CZMG agents are introduced, coordination effects may arise once there is a sufficient number of them, leading to a higher payoff for such agents compared to the DIMG agents even for $m2 = 2$.

8.3.3 Varying the Ratio of CZMG and DIMG Agents in a Population

We now consider the situation when the composition of a population in terms of agents having access to binary and detailed information is varied between the two extreme cases discussed above. It should be intuitively clear that introducing multiple CZMG agents in a population of DIMG agents may lead to the few CZMG agents using the information accessible to them in order to coordinate their actions and thereby increase their payoff. Conversely, introducing multiple DIMG agents in a population of CZMG agents could result in a higher payoff for the few DIMG agents.

Figure 8.4 shows the payoff of the different types of agents, as well as, that of the population as a whole, when the agent composition of the population is altered. Specifically, the fraction of CZMG agents is varied between 0 and 1 keeping the size N of the population constant ($N = 127$ in Fig. 8.4). We observe that CZMG agents having memory length $m1 = 1$ do not have any advantage over the DIMG agents, but as $m1$ is increased they show a relatively better performance for an optimal range of population fraction $f1$ (for simplicity, we keep the memory size of the DIMG agents, $m2$, fixed to 1). For $2 < m1 \leq 0.337 \log_2 N$, we find that there is a peak in the payoff for CZMG agents that occurs for a population fraction between 0 and 1—indicating that having multiple CZMG agents in a population of DIMG agents result in the former having an advantage over the latter under certain conditions. A qualitative argument for the location of this peak in the payoff function can be made as follows. If we ignore for the moment the DIMG agents who are also present in the population, we can consider it as a population comprising only Nf_1 CZMG agents. Given a memory size $m1$, we can determine the optimal population size $N' \sim 2^{m1}/0.337$ at which the collective behavior of the agents achieve maximum efficiency. Thus, if the DIMG agents had no effect on the performance of the CZMG agents, we would have expected the peak at $f_1^* \sim N'/N \sim 2^{m1}/(0.337N)$. However, the interference from these other agents results in the optimal f_1 shifting to lower values. Beyond $m1 = 0.337 \log_2 N$, the payoff for CZMG agents becomes a monotonically increasing function of $f1$.

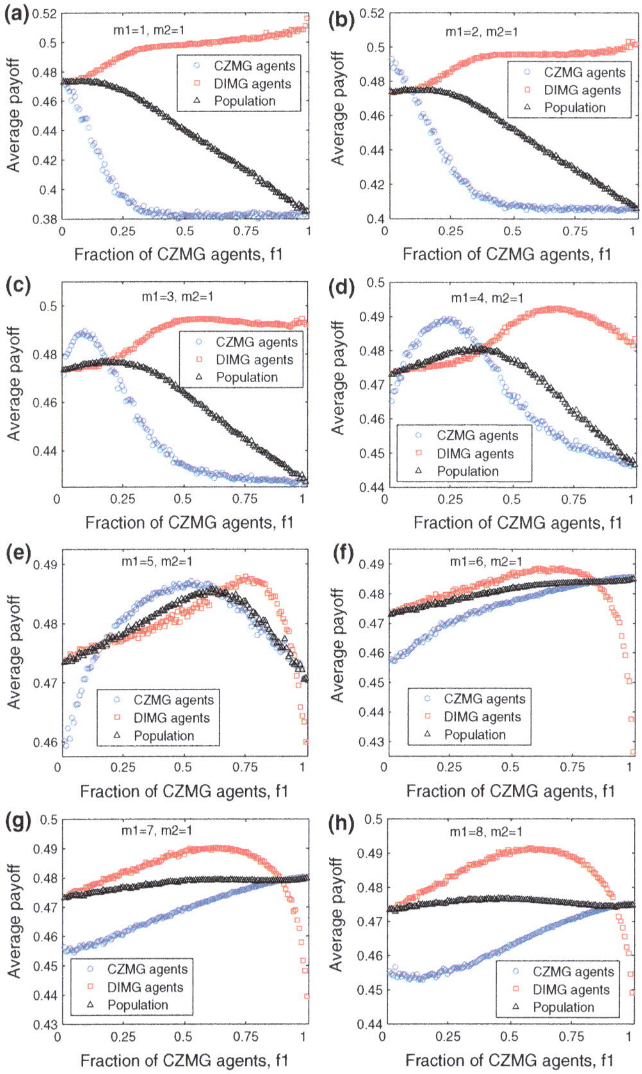

Fig. 8.4 Average payoffs of different types of agents, as well as that of the entire population comprising N agents, shown as a function of the fraction of agents having binary information (CZMG), $f1$, with the remaining agents having detailed information (DIMG). Payoffs are averaged over 10^4 iterations in the steady state and over 250 different realizations with $N = 127$. The different panels show the result of increasing the memory length $m1$ of the CZMG agents by unity from **a** $m1 = 1$ to **h** $m1 = 8$. The memory length of the DIMG agents is fixed at $m2 = 1$

Looking at the DIMG agents, we note that their payoff increases as their population fraction is decreased for very low $m1$, viz., $m1 = 1$ and 2. For larger memory sizes of the CZMG agents, we note that the optimal population fraction of the DIMG agents at which they have maximum payoff occurs between 0 and 1. As $m1$ increases, the advantage of the DIMG agents extend over almost the entire range of $f1$—with CZMG agents having a relative advantage only if they comprise the bulk of the population. In other words, one or a few DIMG agents will not perform very well when facing CZMG agents with sufficiently large memory size $m1$.

8.4 Discussion and Conclusions

In this paper we have considered the effect of information asymmetry between agents on their relative performance in a complex adaptive system. We have used the specific setting of the minority game where agents compete for limited resources, adapting their behavior based on information about past outcomes. By considering heterogeneous composition of agents, who have access to qualitatively different types of information, we have investigated how the granularity of information can affect the payoffs of the agents.

Our results suggest that an agent using information of a particular granularity (i.e., either binary or detailed) may be able to detect any predictable pattern in the outcomes that is generated by the collective behavior of agents who have access to another type of information. This confers an advantage to the former who can then use a strategy which exploits this predictability, providing it with a relatively better payoff. Such an effect is, of course, also dependent on the composition of the population in terms of the different types of agents. Thus, when agents are heterogeneous in terms of the information that is accessible to them—representing a very general situation of information asymmetry observable in almost all real-life situations including markets—it is not the quantity of data, or even the specific nature of the information, available to an agent but rather the ecology of agents with which it interacts that is the key determining factor of its success. Our work implies that simply having access to large volumes of detailed information ("big data") about markets will not translate into higher gains. Indeed, sometimes agents having less data can be more successful—a seemingly paradoxical outcome in terms of mainstream economic thinking about information asymmetry, but which can be understood using the framework discussed here.

In this paper, we have only considered two extreme cases of information granularity, showing that under certain conditions, the coarse-grained data containing only the identity of the option that the minority chooses can be more advantageous than detailed information about how many chose a particular option. However, we can also ask whether there is an optimal level of coarse-graining of information that will confer an advantage in a specific circumstance. We plan to address this issue of how the level of granularity can affect the relative performance of the different types of agents in a future work.

Acknowledgements This work was supported in part by the IMSc Econophysics project (XII Plan) funded by the Department of Atomic Energy, Government of India.

References

Arthur W B (1994) Inductive reasoning and bounded rationality, American Economic Review 84: 406-411

Challet D, Marsili M, Zecchina R (2000) Statistical mechanics of systems with heterogeneous agents: Minority games, Physical Review Letters 84: 1824–1827

Challet D, Marsili M, Zhang Y C (2005) Minority Games: Interacting agents in financial markets. Oxford University Press, Oxford, 2005

Challet D, Zhang Y C (1997) Emergence of cooperation and organization in an evolutionary game, Physica A 246: 407–418

Challet D, Zhang Y C (1998) On the minority game: Analytical and numerical studies, Physica A 256: 514-532

Coolen A C C (2004) The mathematical theory of minority games. Oxford University press, Oxford

Dhar D, Sasidevan V, Chakrabarti B K (2011) Emergent cooperation amongst competing agents in minority games, Physica A 390: 3477–3485.

Hart M, Jefferies P, Hui P M, Johnson N F (2001) Crowd-anticrowd theory of multi-agent market games, European Physical Journal B 20: 547–550

Miller J H, Page S E (2007) Complex adaptive systems: An introduction to computational models of social life. Princeton University Press, Princeton

Moro E (2004) The Minority Game: An introductory guide, in Advances in condensed matter and statistical mechanics (Eds. E Korutcheva and R Cuerno). Nova Science Publishers, New York

Sasidevan V (2016) Effect of detailed information in the minority game: optimality of 2-day memory and enhanced efficiency due to random exogenous data, Journal of Statistical Mechanics: Theory and Experiment 7: 073405

Sasidevan V, Dhar D (2014) Strategy switches and co-action equilibria in a minority game, Physica A 402: 306–317

Simon H (1955) A behavioral model of rational choice, Quarterly Journal of Economics 69: 99–118

Sinha S, Chatterjee A, Chakraborti A, Chakrabarti B K (2010) Econophysics: An introduction. Wiley-VCH, Weinheim

Stiglitz J E (2000) The contributions of the economics of information to twentieth century economics, Quarterly Journal of Economics 115: 1441–1478

Chapter 9
Kolkata Restaurant Problem: Some Further Research Directions

Priyodorshi Banerjee, Manipushpak Mitra and Conan Mukherjee

Abstract In an earlier work on Kolkata paise restaurant problem, Banerjee et al. 2013, we analyzed the cyclically fair norm. We identified conditions under which such a fair societal norm can be sustained as an equilibrium. In this chapter we suggest how the Kolkata restaurant problem can be extended in several directions from purely an economics based modeling perspective.

9.1 Introduction

In the Kolkata restaurant problem (see Chakrabarti et al. 2009; Ghosh et al. 2010), there is a finite set of players who in each period choose to go to any one of the available restaurants. It is assumed that the players have a common ranking of the restaurants. Each restaurant can serve only one customer in any given period. When more than one customer arrives at the same restaurant, only one customer is chosen at random and is served. We can use the tools available in the game theory literature to model the Kolkata restaurant problem of any given day as a one-shot game. Let $N = \{1, \ldots, n\}$ be the set of players ($n < \infty$) and let $V = (V_1, \ldots, V_n) \in \Re^n$ represent the utility (in terms of money) associated with each restaurant which is common to all players. Assume without loss of generality that $0 < V_n \leq \cdots \leq V_1$. Let $S = \{1, \ldots, n\}$ be the (common) strategy space of all players where a typical strategy

P. Banerjee · M. Mitra (✉)
Economic Research Unit, Indian Statistical Institute, Kolkata, India
e-mail: mmitra@isical.ac.in

P. Banerjee
e-mail: banpriyo@isical.ac.in

C. Mukherjee
Department of Econmics, Lund University, Lund, Sweden
e-mail: conanmukherjee@gmail.com

C. Mukherjee
Department of Humanities & Social Sciences, Indian Institute of Technology Bombay, Bombay, India

F. Abergel et al. (eds.), *Econophysics and Sociophysics: Recent Progress and Future Directions*, New Economic Windows,
DOI 10.1007/978-3-319-47705-3_9

$s_i = k$ denotes the strategy that the i-th player goes to the k-th restaurant. The vector $\Pi(s) = (\Pi_1(s), \ldots, \Pi_n(s))$ is the expected payoff vector associated with any strategy combination $s = (s_1, \ldots, s_n) \in S^n$ where player i's payoff is $\Pi_i(s) = V_{s_i}/N_i(s)$ and $N_i(s) = 1 + |\{j \in N \setminus \{i\} \mid s_i = s_j\}|$ is the number of players selecting the same restaurant as that of player i under the strategy combination s. To capture the feature that players prefer getting served in some restaurant to not getting served, we assume that $V_n > V_1/2$. Let $NE(V)$ be the set of all pure strategy Nash equilibria of the one-shot Kolkata restaurant problem. It is easy to check that the set of all pure strategy Nash equilibria of this game, that is, $NE(V) = \{s \in S^n \mid N_i(s) = 1 \; \forall \; i \in N\}$. Let $M(S)$ denote the set of all mixed strategies defined over S. A symmetric mixed strategy Nash equilibrium $\underline{p}^* = (p^*, \ldots, p^*) \in M(S)^n$ where $p^* = (p_1^*, \ldots, p_n^*) \in [0, 1]^n$ with $\sum_{i=1}^{n} p_i^* = 1$ is a solution to the following set of equations: For each $i \in N$, $\sum_{k=0}^{n-1}(1 - p_i^*)^k = [nc(n)]/V_i$ for some constant $c(n)$ which is positive real (see Banerjee et al. 2013). Specifically, for mixed strategy equilibria, the required condition is $\sum_{r=0}^{n-1}\left\{\binom{n-1}{r}(p_i^*)^r(1 - p_i^*)^{n-r-1}[V_i/(r + 1)]\right\} = c(n)$ for all $i \in N$ and after simplification we get $\sum_{k=0}^{n-1}(1 - p_i^*)^k = [(nc(n))/V_i]$ for all $i \in N$. In general, for $n > 3$ such symmetric mixed strategy equilibria always exists (see Becker and Damianov 2006). A general feature of the symmetric mixed strategy equilibria is that $0 < p_n^* \leq \ldots \leq p_1^* < 1$ and $p_1^* \neq p_n^*$.

An allocation of players to restaurants is said to be Pareto efficient if it is not possible to improve the utility of one player without reducing the utility of any other player. The restriction $V_n > V_1/2$ implies that all pure strategy Nash equilibria of the stage game are Pareto efficient. Hence there are exactly $n!$ pure strategy Nash equilibria of this version of the stage game of the Kolkata restaurant problem. If customers are rational, n is small and if customers can mutually interact, then, given that all pure strategy Nash equilibria are Pareto efficient, one can show that it is easy to sustain any pure strategy Nash equilibrium of the stage game of the Kolkata Paise Restaurant problem as a sub-game perfect equilibrium outcome of the Kolkata Paise Restaurant problem without designing any punishment strategy. This is because unilateral deviation here means going to a restaurant where there is already another customer which is payoff reducing. In this context it seems quite unfair to sustain exactly one pure strategy Nash equilibrium of the stage game repeatedly as a sub-game perfect Nash equilibrium of the Kolkata Paise Restaurant problem. This is because in any pure strategy Nash equilibrium of the stage game, the customer going to the first restaurant derives a strictly higher payoff than the customer going to the last restaurant. Instead it seems more natural to sustain the cyclically fair norm where n strategically different Pareto efficient allocations are sequentially sustained in a way such that each customer gets serviced in all the n restaurants exactly once between periods 1 and n and then again the same process is repeated from the $(n + 1)$th period to period $2n$ and so on. Under the large player assumption, a variant of the cyclically fair norm was proposed in Ghosh et al. (2010). However, this type of cyclically fair norm can also be sustained as a sub-game perfect Nash equilibrium because unilateral deviation at any stage means going to a restaurant already occupied by another customer which is always payoff reducing. Therefore, the existing structure of the Kolkata Paise Restaurant problem is such that if the number of customers n is small and if the customers can coordinate their action then the problem becomes uninteresting as there is no need to design punishment strategies to induce customers to remain on the equilibrium path.

9.1.1 An Earlier Work

In an earlier work on Kolkata paise restaurant problem, (Banerjee et al. 2013), we analyzed the cyclically fair norm. We identify conditions under such a fair societal norm can be sustained as an equilibrium. We find that when $V_1 \leq 2V_n$, the cyclically fair norm constitutes a sub-game perfect equilibrium of the repeated game, irrespective of the discount factor. The case $V_1 > 2V_n$ turns out to be far more complex. To keep our analysis tractable, we focus only on the two and three agent cases under this restriction.

For the two agents case, we find that cyclically fair norm constitutes a subgame perfect equilibrium of the repeated game if and only if the agents are sufficiently patient.[1] That is, the social cohesion in providing equal opportunity of having a meal at the better[2] restaurant to both agents requires each agent to have a high acceptance towards delay in consumption at the better restaurant. The exact equilibrium strategy profile σ^c that generates equilibrium play of the cyclically fair norm, is as follows:

(i) Without loss of generality, if period t is odd, then agent i goes to the ith restaurant.
(ii) If period t is even, then for all $i \neq j \in \{1, 2\}$, agent i goes to restaurant j.
(iii) If in any period t, both agents end up at the same restaurant,then both go to restaurant 1 for all times in future.

Note that the third point in the description of σ_i^c is the punishment for deviating from the cyclically fair norm, and it is crucial in sustenance of the equilibrium.

In the three agents case, this punishment behavior of going to the best restaurant constitutes a Nash equilibrium of the stage game if and only if $V_2 < V_1/3$ or $V_3 \leq V_2 = V_1/3$. And hence, if the agents are sufficiently patient, the aforementioned strategy σ^c leads to equilibrium play of the cyclically fair norm. It is easy to verify that if $\max\{V_3, V_1/3\} < V_2 < V_1/2, \sigma^c$ is no longer a sub-game perfect equilibrium. In this case, the equilibrium strategy profile σ^a that generates equilibrium play of cyclically fair norm, when agents are sufficiently patient, is as follows:

---

[1] In particular, the discount factor δ must be in the open interval $\left(\frac{V_1-2V_2}{V_1}, 1\right)$.

[2] Since there are only two agents and two restaurants, notions of better and best are equivalent.

(i) Without loss of generality, at period 1, each agent i goes to restaurant i.
(ii) If agent i goes to restaurant 1 at period $t-1$, then i goes to restaurant 3 at time t.
(iii) If agent i goes to restaurant $k > 1$ at time $t-1$, then i goes to restaurant $k-1$ at time t.
(iv) If any agent i violates either of these conditions (i), (ii) or (iii), leading to a tie at some restaurant in period t, then for all future periods, the other two agents go to restaurant 1.

It can easily be seen that the changed parameter restrictions alter the set of Nash equilibria of the stage game and hence the punishment (upon deviating from the cyclically fair norm) behavior of agents needed to sustain the cyclically fair norm as the equilibrium play needs to be changed accordingly.

Finally, when $V_3 < V_1/2 \leq V_2 \leq V_1$, the strategy profile σ^b, required to sustain the cyclically fair norm as the equilibrium play, when agents are sufficiently patient, becomes quite complicated. In fact, σ^b retains the first three points of σ^a with the fourth point being replaced by the following:

- If there is tie at any restaurant caused by deviation of agent i, then
  - If $i = 1$, then for all future periods, agent 2 goes to restaurant 2 and agent 3 goes to restaurant 1.
  - If $i = 2$, then for all future periods, agent 1 goes to restaurant 1 and agent 3 goes to restaurant 2.
  - If $i = 3$, then for all future periods, agent 1 goes to restaurant 2 and agent 2 goes to restaurant 1.

9.2 Future Research Directions

The Kolkata restaurant problem is an exciting research project that can be extended in several directions. We list a few below.

(A) Our analysis of repeated interaction in the Kolkata restaurant setting relies heavily on each agent being completely informed about past history (that is, all actions taken by all agents at all times in past), in each period. In fact it is essential to devise an equilibrium punishment behavior based of identity of the deviating agent in the three agent case under aforementioned parameter restrictions. However, in practical settings, it may well be that the identity of the deviating agent is not observable to all other conforming agents (see Kandori 2002; McLean et al. 2014). Further, it may well be beyond human capacity to recall all past actions of all agents at all periods of time. Finally, agents may make mistakes in their play of the game, leading to trembles of strategies (see Selten 1975). In all these cases, it would be interesting to study the possibility of sustenance of the cyclically fair norm as equilibrium play.

(B) The Kolkata restaurant problem becomes a game of coordination when $V_1 < 2V_n$. That is, under this parameter restriction, all agents would ideally want to coordinate among themselves and avoid arriving at the same restaurant as another. In fact, the small number of agents may have channels of direct communication through phone or internet connections, and can let each other know about the restaurant that they have chosen to go. However, such information would be unverifiable, as the agents are free to go to a restaurant other than their declared choice. In parlance of game theory, the agents may indulge in cheap talk (see Farrell and Rabin 1996). It would be interesting to find the equilibrium in an extended game that models for such cheap talk, both in one-shot and repeated interaction settings. The solution concepts used in determining equilibrium play could vary from the standard subgame perfection to the recherché forward induction (introduced by Kohlberg and Mertens 1986).

(C) We could conceive of an impartial mediator who addresses the problem coordination using a public randomization device and so, characterize the set of correlated equilibria of the Kolkata restaurant game (see Aumann 1987). It would be of interest to specify these equilibria for both one-shot and repeated interaction cases.

(D) We could also view the Kolkata restaurant problem without the prism of rational (that is, perfectly calculative) human behavior. Indeed, rationality, as espoused in standard game theory, may well be beyond the cognitive capacity of human mind. One alternative to rationality can be found in the biological theory of evolution and evolutionary dynamics. Roughly, it suggests that patterns of human behavior are genetically determined. Each behavioral phenotype in a large population has a degree of success in its interaction with other phenotypes, quantified by its fitness. The number of fitter phenotypes must grow over time, according to a given dynamic of selection, until a stable state is reached. Further, such a stable set must be immune to invasions by genetic mutations. In fact, any mixture of phenotypes is evolutionarily stable if it is the limiting outcome of the dynamics of selection from any arbitrary mixture of phenotypes in the population (see Taylor and Jonker 1978). It would be of interest to characterize the set evolutionarily stable strategies in the Kolkata restaurant problem with large (finite or infinite) populations.

(E) The basic question of whether the cyclically fair norm can be sustained through decentralized repeated interaction may be investigable in the laboratory using volunteer subjects. Such investigations may help address inquiries as to when the norm is sustainable and what explains failures, if any, of players to achieve coordination. Many of the directions listed above may also lend themselves to experimental investigation, which can provide understanding as to how for instance information on past actions and outcomes, or cheap talk and communication, or public mediation, or the presence of behavioral types, can impact tendencies toward norm formation and maintenance. An laboratory experimental literature has recently emerged studying mainly financial market phenomena such as excess and coordinated or synchronized trading, herd behavior, volatility, bubbles and crashes etc. using minority games as an underlying model of interaction. It is natural to wonder if available results are robust to generalizations to the minority game as represented by the KPR problem. Additionally, existing papers have restricted themselves to anonymous or randomly matched interaction: a key question in this context is whether results extend when players interact repeatedly.

References

R. J. Aumann. Correlated equilibrium as an expression of bayesian rationality. *Econometrica*, 55:1–18, 1987.

P. Banerjee, M. Mitra, and C. Mukherjee. *Econophysics of Systemic Risk and Network Dynamics*, chapter Kolkata Paise Restaurant Problem and the Cyclically Fair Norm, pages 201–216. Springer Milan, 2013.

J. G. Becker and D. S. Damianov. On the existence of symmetric mixed strategy equilibria. *Economics Letters*, 90:84–87, 2006.

A. S. Chakrabarti, B. K. Chakrabarti, A. Chatterjee, and M. Mitra. The kolkata paise restaurant problem and resource utilization. *Physica A: Statistical Mechanics and its Applications*, 388:2420–2426, 2009.

J. Farrell and M. Rabin. Cheap talk. *The Journal of Economic Perspectives*, 10:103–118, 1996.

A. Ghosh, A. Chatterjee, M. Mitra, and B. K. Chakrabarti. Statistics of the kolkata paise restaurant problem. *New Journal of Physics*, 12:075033, 2010.

M. Kandori. Introduction to repeated games with private monitoring. *Journal of Economic Theory*, 102:1–15, 2002.

E. Kohlberg and J-F. Mertens. On strategic stability of equilibria. *Econometrica*, 54:1003–1037, 1986.

R. McLean, I. Obara, and A. Postlewaite. Robustness of public equilibria in repeated games with private monitoring. *Journal of Economic Theory*, 153:191–212, 2014.

R. Selten. Reexamination of the perfectness concept for equilibrium points in extensive games. *International Journal of Game Theory*, 4:25–55, 1975.

P. D. Taylor and L. B. Jonker. Evolutionary stable strategies and game dynamics. *Mathematical Biosciences*, 40:145–156, 1978.

Chapter 10
Reaction-Diffusion Equations with Applications to Economic Systems

Srinjoy Ganguly, Upasana Neogi, Anindya S. Chakrabarti and Anirban Chakraborti

Abstract In this article, we discuss reaction-diffusion equations and some potential applications to economic phenomena. Such equations are useful for capturing non-linear coupled evolution of multiple quantities and they show endogenous oscillatory behavior as well as non-convergence to a constant equilibrium state. We model technological competition and spill-over of productivity shocks across countries using simple variants of reaction equations of the Lotka-Volterra type. Coupled with standard real business cycle models for individual countries, this gives rise to non-trivial lag-lead structure in the time-series properties of the macroeconomic variables across countries. We show that this simple model captures a number of properties of the real data.

10.1 Introduction

Macroeconomic time-series often show a clear lag-lead structure. Specifically in the case of macroeconomic booms and busts, there is clear signature of spill-over effects from one country to another (a popular catch-phrase is that when one country catches a cold, its economic partners sneeze) and such spill-over effects are often not linear. In the present world, nonlinearity arises from a high degree of interconnection

S. Ganguly
Indian Institute of Management, Vastrapur, Ahmedabad 380015, India
e-mail: p14srinjoyg@iima.ac.in

U. Neogi · A. Chakraborti
School of Computational and Integrative Sciences, Jawaharlal Nehru University, New Delhi 110067, India
e-mail: upowers19@gmail.com

A. Chakraborti
e-mail: anirban@jnu.ac.in

A.S. Chakrabarti (✉)
Economics Area Indian Institute of Management, Vastrapur, Ahmedabad 380015, India
e-mail: anindyac@iimahd.ernet.in

F. Abergel et al. (eds.), *Econophysics and Sociophysics: Recent Progress and Future Directions*, New Economic Windows,
DOI 10.1007/978-3-319-47705-3_10

across multiple economic entities. Delligatti et al. (1998) empirically documents that country-specific GDP can be described well by nonlinear dynamics. International business synchronization literature is fairly large. Interested readers are referred to Bordo and Helbing (2010) (and references therein) for a comprehensive econometric analysis.

In this paper, our goal is to present a very simple idea to capture such nonlinear dependence of macroeconomic quantities. Broadly, the idea is that in an economy with multiple constituent countries, each country can be described by a standard business cycle model and the only linkage across these countries are given by technology flows which evolve jointly following a set of coupled nonlinear equations. Under certain conditions imposed on the parameter values, technology flow shows limit cycles which in turn causes fluctuations in the macroeconomic variables that captures a clear lag-lead structure in time-series behavior as well as endogenous business cycles. Even though the present framework is arguably mechanical in its approach to generate the co-evolution of business cycles, it is useful for a parsimonious description.

In the following, we first describe the mathematical properties of the reaction-diffusion systems that forms the basis of description of the linkages. Then we discuss a particular instance of it, which maps directly into Lotka-Volterra type interactive systems and characterize its phase-diagram. Then we describe an application to study spill-over effects on economic entities.

10.2 Mathematical Description

Reaction-diffusion equations are mathematical models applied in a wide variety of subjects. The general mathematical formalism has been applied to biological, physical and economic systems among others. These systems are expressed by partial differential equation that are semi-linear and parabolic in nature. The most common application of reaction-diffusion equation is in chemical reaction in which the constituents are transformed locally into each other and transported over a surface in space through diffusion (see e.g. Reaction-diffusion system 2016 for a general description).

The standard form of reaction-diffusion equation is given by

$$\frac{\partial u}{\partial t} = D\nabla^2 u + R(u), \tag{10.1}$$

where $u(x, t)$ represents the vector function to be calculated, D is the diagonal matrix representing the diffusion coefficients and R is the function that describes the local reaction. The reaction-diffusion equations are specified by the differential equations, initial conditions and boundary conditions. But there will be some special conditions at the boundary where the differential equation does not apply. For example, in chemical systems, the walls of a container are impermeable to the chemicals. Hence, a certain condition is applied which defines that the chemicals cannot leak through the walls (Marc Roussel 2005). Such boundary conditions are called no-flux boundary conditions and mathematically represented as

$$\bar{J}.\hat{n} = 0, \tag{10.2}$$

where $\bar{J}$ is the flux and $\hat{n}$ is a normal vector to the boundary.

10.2.1 Reaction-Diffusion Equation (One Dimension)

The simplest form of reaction-diffusion equation for one component (i.e. one dimensional) is

$$\frac{\partial u}{\partial t} = D\frac{\partial^2 u}{\partial^2 t} + R(u), \tag{10.3}$$

Equation 10.3 is also referred as the Kolmogorov-Petrovsky-Piskounov equation (KPP equation). KPP equations are mainly used in the case of spatial evolution of a state while propagation in a homogeneous medium. If the reaction term is omitted, the equation represents the law for diffusion which is the Fick's second law (Reaction-diffusion system 2016; Reardon and Novikov 2016).

10.2.2 Reaction-Diffusion Equation (Two Dimensions)

Two-component or two-dimensional systems are largely used in the ecological problems for prey-predator interaction and in chemistry where new substances are produced from reaction of different substances (Junping Shi 2004). If u(x, t)and v(x, t) are assumed as the density functions of two populations, then we can write the corresponding equations as,

$$\begin{aligned}\frac{\partial u}{\partial t} &= D_u\frac{\partial^2 u}{\partial^2 t} + F(u, v),\\ \frac{\partial v}{\partial t} &= D_v\frac{\partial^2 v}{\partial^2 t} + G(u, v),\end{aligned} \tag{10.4}$$

where the coefficients (D_u, D_v) are the diffusion constants and the additive terms $(F(u, v), G(u, v))$ represent the reaction functions (Junping Shi 2004).

In 1937, Fisher introduced the idea of reaction-diffusion models for the spatial dispersion of a particular gene. In 1952, Alan Turing first proposed an idea that in presence of diffusion, a stable state can potentially become unstable. He suggested that the linearly stable uniform steady state with two or more components can destabilize in presence of diffusion and formation of spatial inhomogeneous patterns take place through bifurcation (Reaction-diffusion system 2016; Junping Shi 2004; Robert Stephen Cantrell and Chris Cosner 2003).

For example, a spatio-temporal interacting species model can self-organize and show different kinds of patterns, both stationary and non-stationary. The former types are associated with spatial distribution of the species in a non-constant steady-state. These can be of multiple types: examples include cold-spots and hot-spots, stripe/labyrinthine or a mix of the two. Turing (or Turing-Hopf)-domain in the parameter space is typically the domain where such patterns materialize. On the other hand, non-stationary patterns do not reach a steady-state which is non-constant and continuously evolve over time. Such patterns can take the form of wave-like periodic, spiral or even chaotic patterns.

For a much more detailed discussion on this topic, interested readers are referred to Sirohi et al. (2015). We do not pursue this discussion further as it lies outside the scope of the present article.

10.3 Lotka-Volterra (LV) Model: Predator-Prey Interactions

For a very long time, researchers have investigated the famous prey-predator model, i.e., Lotka-Volterra system of equations for analyzing endogenous oscillatory behavior of coupled nonlinear equations. Sirohi et al. (2015) recently introduced environmental noise in such a model and studied the corresponding effects on spatio-temporal pattern formation. Due to the addition of environmental noise, random fluctuations have been observed for the fundamental variables characterizing the predator-prey system, viz. carrying capacity, intensity of intra- and inter-species competition rates, birth and death rates, and predation rates. In particular, they study a simple predator-prey model with "ratio-dependent functional response, density dependent death rate of predator, self-diffusion terms corresponding to the random movement of the individuals within two dimension, in addition with the influence of small amplitude heterogeneous perturbations to the linear intrinsic growth rates".

Below we summarize the formulation in Sirohi et al. (2015) following their notations, in order to show that the LV mechanism can be augmented by stochastic noise terms, still retaining non-trivial nonlinear behavior. We will use a modification of such a model while setting up the economic application in the next section. The nonlinear coupled partial differential equations representing the prey-predator interaction, are given as

$$\frac{\partial u}{\partial t} = u(1-u) - \frac{\alpha uv}{u+v} + \nabla^2 u \equiv f(u,v) + \nabla^2 u,$$
$$\frac{\partial v}{\partial t} = \frac{\beta uv}{u+v} - \gamma v - \delta v^2 + d\nabla^2 v \equiv g(u,v) + \nabla^2 v \qquad (10.5)$$

where $u \equiv u(x, y, t)$ and $v \equiv v(x, y, t)$ denote the population densities of prey and predator respectively at a generic time point (t) and within $\Omega \subset R^2$ with boundary $\partial\Omega$ in the real space. Initial conditions have been set to

$$u(x, y, 0) > 0, \quad v(x, y, 0) > 0 \;\; \forall (x, y) \in \Omega \tag{10.6}$$

The no-flux boundary conditions are represented in the following way,

$$\frac{\partial u}{\partial \nu} = \frac{\partial v}{\partial \nu} = 0 \tag{10.7}$$

$\forall (x, y) \in \partial\Omega$ and positive time points, where ν is the unit normal vector drawn outward on $\partial\Omega$, with scaler parameters α, β, γ and δ.

The magnitude of the parameters used in this model determine whether the Turing patterns will exist or not. In Sirohi et al. (2015), authors have considered α and d as the bifurcation parameters for constructing the Turing bifurcation diagram. The Turing bifurcation diagram has been presented in αd-parametric plane (Fig. 10.1; Sirohi et al. 2015). Value of the parameters in the given bifurcation diagram are $\beta = 1$, $\gamma = 0.6$ and $\delta = 0.1$. In the present formulation, α and d can be controlled to produce spatio-temporal patterns of different kinds. The curves shown in the figure are the Turing-bifurcation curve, temporal Hopf-bifurcation curve and temporal homoclinic bifurcation curve which have been marked as blue curve, red-dashed line and black-dotted line respectively. The equilibrium point E_* destabilize at $\alpha_h = 2.01$ which gives the Hopf-bifurcation curve. The condition for stability of the equilibrium point is $\alpha < \alpha_h$. The region lying above the Turing bifurcation curve is the Turing instability region which is divided into two parts by the Hopf-bifurcation curve. Turing-Hopf domain with unstable temporal and spatio-temporal patterns lie in the region where $\alpha > \alpha_h$. See Sirohi et al. (2015) for further numerical details.

Equation 10.5 produces multiple interesting patterns which can be seen from Fig. 10.1. Within the Turing domain, cold-spots and a mix of spots and stripes materialize. Within the Turing-Hopf domain, a mix of spot-stripe, labyrinthine as well as chaotic patterns materialize. The patterns obtained in Fig. 10.1 have been marked with four different symbols depending on values of α and d with the details in the caption.

The previous model (described in Eq. 10.5) can be augmented by introducing uncorrelated multiplicative white noise terms. The new model is described as

$$\frac{\partial u}{\partial t} = u(1-u) - \frac{\alpha u v}{u+v} + \nabla^2 u + \sigma_1 u \xi_1(t, x, y), \tag{10.8}$$

$$\frac{\partial v}{\partial t} = \frac{\beta u v}{u+v} - \gamma v - \delta v^2 + \nabla^2 v + \sigma_2 u \xi_2(t, x, y), \tag{10.9}$$

where $\xi_1(t, x, y)$ and $\xi_2(t, x, y)$ are i.i.d. noise (a less strict definition would be temporally as well as spatially uncorrelated noise terms with normal distribution) with mean 0,

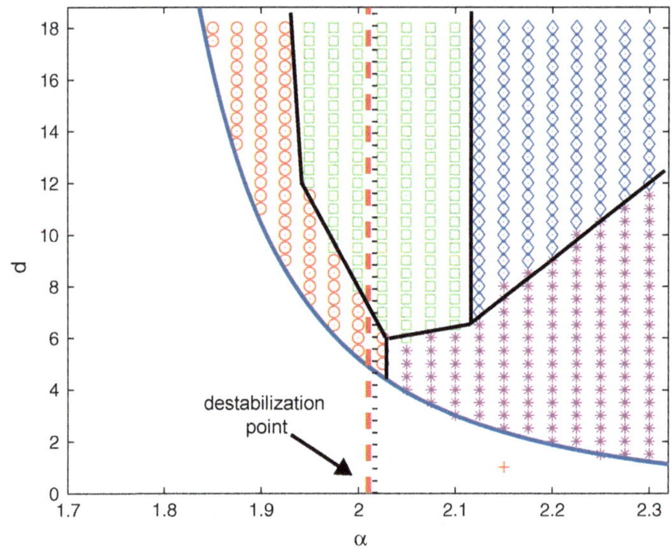

Fig. 10.1 Multiple types of spatial patterns can be observed for different parameter values within the Turing instability region in the parameter space. Four different colored symbols mark the relevant regions in the following way: ∘ cold-spot, □ mixture of spot-stripe, ⋄ labyrinthine, $*$ chaotic and $+$ interacting spiral. Turing bifurcation curve is shown by *blue curve*, Hopf-bifurcation curve by *red dashed line* and temporal homoclinic bifurcation curve by *black dotted curve*. Adapted from Anuj Kumar Sirohi Malay Banerjee and Anirban Chakraborti (2015)

$$E\left(\xi_1\left(t, x, y\right)\right) = E\left(\xi_2\left(t, x, y\right)\right) = 0, \tag{10.10}$$

and σ_1 and σ_2 parameterizes the environmental effects.

Sirohi et al. (2015) conducts simulations of the system describes above and documents that small magnitudes of noise intensities do not have any substantial effect on the spatiotemporal pattern formation apart from the fact that the time taken to reach the stationary pattern increases, which is expected. On the other hand, addition of small noise increases the irregularity within the non-stationary zone. They concluded noise and statistical interaction have vital roles in determining the distribution of species, even when environmental conditions are unfavourable.

To summarize the material discussed above, we have seen that simple LV systems generate many intricate patterns. In general not everything would be useful for economic applications. We borrow two well known insights from this literature. One, nonlinear dependence may lead to endogenous dynamics in the form of perpetual oscillation. That can, at least in principle, be useful to describe evolution of fluctuating macroeconomic variables. Second, it is related to the idea that a dynamic system may not reach a constant equilibrium after all. This has been a point of discussion in multiple occasions among physicists and economists (Sinha et al. 2010). Most of the standard macroeconomic theory is driven by dynamical theories built around a constant equilibrium implying unless a shock hits the economy (productivity, monetary, fiscal policy etc.) it will not show any adjustment. However, it may seem to be somewhat unrealistic feature of such models.

Below we describe a model which is partly borrowed from the economic literature and partly depends on the LV type mechanisms described above. The idea is that given a shock process, an economic model describes evolution of macroeconomic quantities reasonably well. However, where that shock is coming from typically remains unanswered. We opine that in a multi-country context, there can a be a leader-follower relationship across countries that mimics the dynamics described above. The value addition of that approach is that the dynamics of the shock process can be totally endogenous and non-convergent to a constant equilibrium. Correspondingly, macro variables will also show a lag-lead structure as we describe below.

10.4 LV Equations: Modeling Diffusion of Technology

The most commonplace instance of a complex adaptive system would be a decentralized market economy (Sinha et al. 2010). Characterized by adaptive agents regularly partaking in multi-level interactions, such dynamic systems essentially bear witness to macroeconomic regularities, which persist in the form of recurrent causal chains binding individual entities. Furthermore, such a system posits a nuanced symmetric feedback between its micro-structure and macro-structure components, thereby obfuscating the quantitative modeling of the same. Econophysics (Sinha et al. 2010) has traditionally attempted to provide a fundamental framework to understand emergence of patterns in presence of feedback. However, both of the aforementioned problems are indeed challenging because whenever an empirical regularity is established, further work leverages upon the same, thereby undoing its efficacy. In fact, this phenomenon becomes especially pronounced in the realm of economics at the macro level. Given the wide acceptance of the idea that economies cannot exist in silos, it is quite evident that the growth prediction for each country in the current globalized world would have a non-linear dependence upon the situation prevalent in the rest of the world. In this regard, the global recession of the last decade has affected a paradigmatic shift in the modeling of international interactions, which goes beyond linear cause-effect relationships. The said change is predominantly inclined towards the analysis of the complex network of interactions occurring between the heterogeneous economic entities (viz. countries) involved in a trade/capital/technology flow. This non-linear dependence captures the basic property of spill-over effects.

In this part of the paper, we would apply the previously discussed tools to model the complex interdependence exhibited by macroeconomic variables across countries in terms of inter-temporal technological diffusions across countries. It is assumed that, upon considering the international technological frontier as the background, few countries emerge as predators while the others as preys. This means that while the countries belonging to the latter category invest their resources to develop newer technology, those belonging to the latter simply copy the same without undertaking original innovation. This phenomenon, by diminishing the advantage of usage, ultimately results in an adverse impact upon the relative productivity of prey countries. In this regard, one ought to note that the followers (predators) may copy/borrow the technological innovations referenced herein from the leaders (prey) only after the technological boom has occurred in the former, i.e. the predators are separated from the prey by a time lag. As can be evinced from the problem characteristics discussed thus far, it is quite evident that the required model is one which should be able to not only account for the non-linear interdependence of macroeconomic variables but also model predator-prey interactions. This report relies upon the famous Lotka-Volterra model (referred to as the LV model, henceforth), derived from the discipline of theoretical ecology, to achieve the stated objectives. In the past, the LV model has found extensive application in the modeling of financial markets (Solomon 2000), i.e. primarily heterogeneous entities, which is in perfect resonance with the theme of the proposed work. Furthermore, the LV models advantage lies in two avenues. First, it posits an endogenous source of non-linearity thereby nullifying the need for any external non-linearity which might not be analytic. Secondly, the model posits a time lag between the predator and the prey populations which, as discussed earlier, is reminiscent of economic phenomena.

For an empirical motivation of the model discussed below, see Fig. 10.2. We have detrended the yearly per capita GDP series with HP filter for two countries (USA and Italy; data obtained from OECD database). The cyclical components roughly show a lagged behavior. This is precisely the feature we want to model.

10.4.1 A Two-Country Example

Much of the discussion on the model set-up depends on the materials described in Chakrabarti (2016). A miniature two-country example has been used to show how one may derive cross-correlation and auto-correlation patterns from the same. In essence, while the small scale real business cycle model has been used to relate the

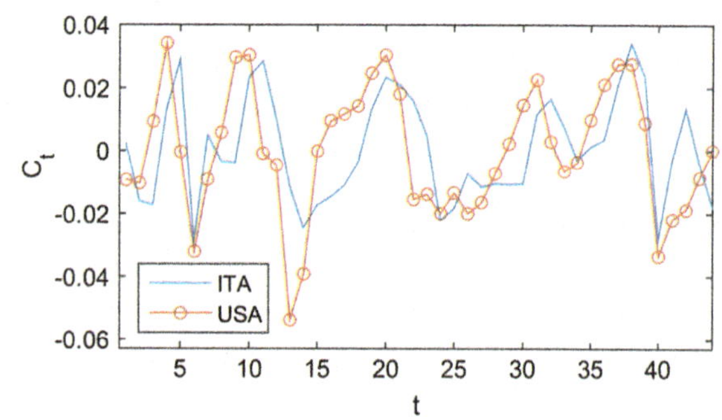

Fig. 10.2 Cyclical components of the yearly per capita GDP series for Italy and USA from 1970–2015. Data has been detrended with HP filter

major variables, the LV model serves to describe the evolution of the technology frontier. In order to leverage upon the leader-follower model, we may adopt a rather simplistic way for gauging a countries technological prowess, viz. the number of blue-prints generated by that nation. Quite intuitively, a country generating more number of blue-prints (a leader/prey) would also enable its inhabitants to wield a higher rate of productivity. On the contrary, while a leader invests resources to generate novel blue-prints, another country (a follower/predator) may simply exhibit parasitic behavior in merely copying/reverse- engineering those blue-prints, thereby following the leader on the technology frontier. Variants of the real business cycle model have been employed extensively in the past, both in continuous and discrete forms. The basic assumptions underlying the formulation of such a miniature model may be summarized as:

- Two distinct economies are considered wherein the leader and the follower are referred to by L and F respectively, wherein ceteris paribus both the economies are endowed with an equal potential to innovate.
- Both the economies are populated by a unit mass of households. Output is produced using a combination of labor and capital. The output so produced may either be immediately consumed or saved and invested.
- We assume single good economies. Furthermore, at the micro as well as the macro level, the fundamental aim of each entity in the system is to maximize his/her/their objective (profit/utility) function.

10.4.1.1 Description of the Economy

The economies have textbook descriptions. Utility function in the j-th country (where $j \in \{L, F\}$):

$$U^j = \sum_{t=0}^{\infty} \beta^t \left(\ln C_t^j + \alpha \ln(1 - L_t^j)\right), \tag{10.11}$$

where C_t^j and L_t^j denotes consumption and labor respectively. The production function is defined by

$$Y_t^j = z_t^j (K_t^j)^\theta (L_t^j)^{1-\theta}, \tag{10.12}$$

where capital is denoted by K_t^j, labor by L_t^j and technology by z_t^j. Capital accumulation occurs following the standard description,

$$K_{t+1}^j = (1 - \delta) K_t^j + I_t^j, \tag{10.13}$$

where δ is the rate of depreciation and I_t is the investment. There is a resource constraint that

$$C_t^j + I_t^j \leq Y_t^j. \tag{10.14}$$

The evolution of the technology term z_t^j introduces the linkage across economies. For this type of models, z_t^j represents a technological shock process orthogonal to the other macroeconomic fundamentals of the economy. Here, we impose a nonlinear coupling of evolution of technology to study the impacts on other variables.

10.4.1.2 Linkages Across Countries

The evolution of the technology is given by a general form (see Chakrabarti 2016 for details)

$$Z(t+1) = \Gamma(Z(t)), \tag{10.15}$$

where $\Gamma()$ defines the interaction terms. A specific description would be:

$$\begin{aligned} \frac{dZ^L}{dt} &= aZ^L - bZ^L Z^F, \\ \frac{dZ^F}{dt} &= -cZ^L + dZ^L Z^F. \end{aligned} \tag{10.16}$$

10.4.1.3 Discretization

Since we are attempting to model the economy in discrete time, we cannot directly work with the continuous time-path of technology. Instead, we will focus on a stochastically sampled sequence of the time-path generated by the set of equations. Let us assume that the observed discrete sequence is denoted by $\{S_\tau\} = \{Z^L(\tau), Z^F(\tau)\}$ where $\tau = 1, 2, 3, \ldots$ Since the variance of the actual economic fluctuations can be found from the data, we need the technology process to possess tunable variance. Therefore, we transform the original series in the following way: the shock process as

$$z^j(t) = \frac{1}{1 + e^{-\kappa S^j(t)}} \quad j \in \{L, F\}, \tag{10.17}$$

where κ is a constant.

Figure 10.3 describes the essential dynamic behavior of the system. Under standard parameter values (we assume 1, -1, 1 and -1 in Eq. 10.16 for a, b, c, d resp.), the limit cycles are evident. Then we carry out the transformation to tune volatility of the series with Eq. 10.17. Upon random sampling, the resultant time-series has non-trivial properties. See Chakrabarti (2016) for further details.

10.4.1.4 Mechanism

Note that the model economy is fully described by Eqs. 10.11, 10.12, 10.13 and 10.14. In particular, Eq. 10.12 contains the only exogenous variable z. The rest of

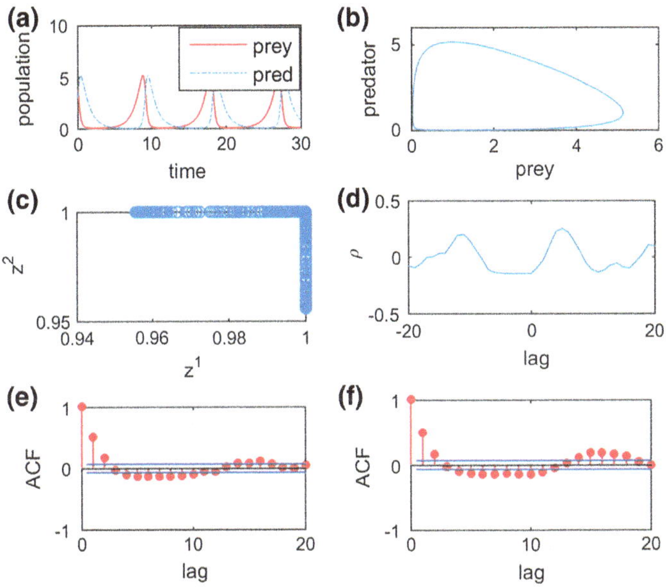

Fig. 10.3 Description of the time-series behavior Lotka-Volterra model: **a** time-domain description of a two species LV model, **b** phase diagram for the same, **c** phase diagram after applying the required transformation (Eq. 10.17), **d** cross-correlation of the time-series constructed by randomly sampling of the predator and prey series, **e**–**f** autocorrelation functions of the same

the variables can be determined upon description of z. We posit that the evolution of this variable is given by the transformed version of the LV equations viz. Eq. 10.17.

Because of the underlying LV mechanism, the technology variable z would show endogenous oscillatory behavior. Couple with the description of the evolution of the macro variables, we can solve for each of the variables, all of which in turn would show endogenous dynamics along with coupling behavior.

10.4.2 A Generalized Framework

The system of two-country LV equations may be extended to a general form for an N-country scenario (with each country indexed by $i \in \{1, 2, ..., N\}$):

$$\frac{dz^i}{dt} = \mu_{ii} z^i + \sum_{j \neq i} \mu_{ij} z^i z^j \quad \forall\, i \in N. \tag{10.18}$$

This set of equations represent the most fundamental form of the LV model, which has a number of variants depending on the specific modeling requirements. The following equation represents a very generalized form of the interaction equations along with a squared term:

$$\frac{dz^i}{dt} = \sum_j \mu_{ij1}\bar{z}^j + \mu_{ii2}(\bar{z}^i)^2 + \sum_j \mu_{ij}\bar{z}^i\bar{z}^j \quad \forall\, i \in N. \tag{10.19}$$

As proposed in Wu and Wang 2011, the LV models in-sample predictions offer differ significantly from the actual time-series owing to the lack of any implicit/explicit time-smoothening per se in the model. In order to alleviate this bottleneck, one may leverage upon Grey modeling (Wu and Wang 2011) to obtain several variants of grey Lotka-Volterra (GLV) models.

In order to move towards a method for estimating the parameters, we need to discretize the process. A simple way to do that is to assume an approximation that the left hand side of Eq. 10.19 is $z^i(t+1) - z^i(t)$ (Wu and Wang 2011). For the proposed methodology, one can construct a backward-looking weighted average of the past values as the basis for predicting the future,

$$\bar{z}^i(\tau) = \frac{\sum_{n=0}^{w} k^n z^i(\tau - k)}{\sum_{n=0}^{w} k^n}. \tag{10.20}$$

10.4.2.1 Estimation Techniques

One of the aims of the given exercise is to estimate the parameters in Eq. 10.19 using multiple time-series data for macroeconomic quantities. Thus, the next step would essentially entail arriving at appropriate equation parameter values so as to ensure that the in-sample error factor (η_i) is minimized.

After estimation, the model should essentially predict the future behavior of the time-series. Below, we first describe one possible way to estimate the parameters. Numerically, we see that prediction for the next time instant ($\hat{z}(t+1)$) by leveraging upon immediate past data has a good in-sample fit. Unfortunately, the out-of-sample fits are very bad for reasons described towards the end of this section.

The proposed scheme adopts a convex optimization approach to achieve the said objective, i.e. it leverages upon first-order conditions to solve the set of parameter values. The in-sample error is given by

$$\eta_i = \sum_t \left(\hat{z}^i(t) - z^i(t)\right)^2 \tag{10.21}$$

where $z^i(t)$ is the actual observed series. The first order conditions will be given by

$$\frac{\delta \eta_i}{\delta \mu_k} = 0, \tag{10.22}$$

where μ_k are the parameters describing the generalized LV system. Required second-order condition would be that the matrix of the second derivative would be negative definite. These are fairly standard conditions for multi-valued optimization and hence we skip elaborating on the same.

10.4.3 Lack of Parsimony for Prediction

The present approach has good in-sample fit (see Ganguly 2016 for some preliminary results). However, the out-of-sample fit has a number of problems. One, the generalized version given by Eq. 10.19 is not guaranteed to converge and in fact, in most of the cases checked numerically, it actually diverges. Thus predictions beyond a couple of time-points are not particularly reliable. One potential problem is the approximation that the time derivative is given by $z^i(t+1) - z^i(t)$ although it is not clear what else can be used instead. Second, much of the goodness of fit even with the in-sample data is driven by overfitting as there are $N \times (2N+1)$ number of parameters that can be calibrated. Hence, the generalized version, even though it nests many dynamical systems upon careful selection of parameters, is not the best candidate for prediction purpose.

10.4.4 Characterizing the Spill-Over Effects

The basic LV framework cannot describe the spill-over effects. The basic problem is that if we define the z^i term to be the deviations from the steady state, then the spill-over effect is always zero starting from any $z^j = 0$. Note that in the basic LV model, the interaction terms are multiplicative, so even if the deviation for the i-th country can be non-zero by a stochastic shock, due to the multiplicative characteristics, the product necessarily has to be zero. Hence, LV mechanism by itself is not capable of generating spill-over effects in terms of, for example, cross-country impulse response functions. However, the augmented version Eq. 10.19 contains autoregression terms $(\sum_j \mu_{ij1}\bar{z}_j)$ in levels without any multiplicative factors and hence, this will allow characterization of spill-over effects.

10.5 Summary and Outlook

We have presented a general model for reaction-diffusion equations and discussed some specific applications to generate lag-lead structure of competing economic entities. Further developments on usage of nonlinear interactive systems to characterize economic time-series properties would be interesting additions to the literature.

Finally, one important issue regarding nomenclature is that we are using the words predator and prey only for descriptive purpose. Such usage should not be taken to be implying anything other than the specific technical meaning assigned to them by the standard LV mechanism. An economic interpretation can be immediately given by considering a leader-follower duo competing on the technology frontier.

Acknowledgements A.C. acknowledges financial support from the Institutional Research Funding IUT(IUT39-1) of the Estonian Ministry of Education and Research, and grant number BT/BI/03/004/2003(C) of Government of India, Ministry of Science and Technology, Department of Biotechnology, Bioinformatics Division. U.N. and A.C. acknowledge support from the University of Potential Excellence-II grant (Project ID-47) of the Jawaharlal Nehru University, New Delhi, India.

References

M. Bordo and T. Helbing. International business cycle synchronization in historical perspective. *NBER working paper series*, 2010.

R. S Cantrell and C. Cosner. Spatial ecology via reaction-diffusion equations. Wiley and Sons Ltd., 2003.

A. Chakrabarti. Stochastic lotka-volterra equations: A model of lagged diffusion of technology in an interconnected world. *Physica A*, 442:214–223, 2016.

D. Delligatti, M. Gallegati, and D. Mignacca. Nonlinear dynamics and european gnp data. *Studies in Nonlinear Dynamics and Econometrics*, 3(1):43–59, 1998.

S. Ganguly. Inter-temporal correlation structure of cross-country technology diffusion: An application of the generalized lotka- volterra model. *SSRN-id2699934*, 2016.

Reaction-diffusion system, https://en.wikipedia.org/wiki/reaction. 2016

A. Reardon and A. Novikov. Front propagation in reaction-diffusion equations, https://www.math.psu.edu/wim/reports/amberreardon.pdf. 2016

M. R. Roussel. Reaction-diffusion equations, http://people.uleth.ca/~roussel/nld/turing.pdf. 2005.

J. Shi. Reaction diffusion systems and pattern formation, http://www.resnet.wm.edu/~jxshix/math490/lecture-chap5.pdf. 2004.

S. Sinha, A. Chatterjee, A. Chakraborti, and B. K. Chakrabarti. Econophysics: An introduction. *Wiley-VCH*, 2010.

A. K. Sirohi, M. Banerjee, and A. Chakraborti. Spatiotemporal pattern formation in a prey-predator model under environmental driving forces. *Journal of Physics, Conference Series*, page 012004, 2015.

S. Solomon. Generalized lotka-volterra (glv) models of stock markets. *Advances in Complex Systems*, 3(301), 2000.

L. Wu and Y. Wang. Estimation the parameters of lotka-volterra model based on grey direct modelling method and its application. *Expert Systems with Applications*, 38:6412–6416, 2011.

Part II
Sociophysics

Chapter 11
Kinetic Exchange Models as *D* Dimensional Systems: A Comparison of Different Approaches

Marco Patriarca, Els Heinsalu, Amrita Singh and Anirban Chakraborti

Abstract The Kinetic Exchange Models represent a charming topic in both interdisciplinary physics, e.g. in the study of economy models and opinion dynamics, as well as in condensed matter physics, where they represent a simple but effective kinetic model of perfect gas, with the peculiar feature that the dimension D of the system is a real variable which can be tuned continuously. Here we study kinetic models of energy exchange between particles of a perfect gas in D dimensions and discuss their relaxation toward the canonical equilibrium characterized by the energy distribution in D dimensions ($D \in \mathbb{R}$), comparing various theoretical approaches with results from numerical simulations.

11.1 Introduction

Kinetic Exchange Models (KEMs) have attracted considerable attention not only in the interdisciplinary physics, whether opinion dynamics or the studies of wealth exchange models, but also in condensed matter physics as in the case of prototypical and general systems of units exchanging energy (Patriarca and Chakraborti 2013). A noteworthy feature of KEMs is that a suitable tuning of some parameters regulating the energy exchange in the basic homogeneous versions leads to a situation where

M. Patriarca (✉) · E. Heinsalu
NICPB–National Institute of Chemical Physics and Biophysics,
Rävala 10, 10143 Tallinn, Estonia
e-mail: marco.patriarca@kbfi.ee

E. Heinsalu
e-mail: els.heinsalu@kbfi.ee

A. Singh · A. Chakraborti
SCIS–School of Computational & Integrative Sciences,
Jawaharlal Nehru University, New Delhi 110067, India
e-mail: amritasingh@jnu.ac.in

A. Chakraborti
e-mail: anirban@jnu.ac.in

F. Abergel et al. (eds.), *Econophysics and Sociophysics: Recent Progress and Future Directions*, New Economic Windows,
DOI 10.1007/978-3-319-47705-3_11

the system relaxes toward a Boltzmann canonical energy distribution characterized by an arbitrary dimension D. It is not that D can assume only a positive integer value. In fact, it is a real variable that can assume any value greater than or equal to 1. In this contribution we discuss a basic version of KEM in D dimensions using different theoretical approaches, including numerical simulations of a perfect gas in D dimensions. This provides a historical overview of KEMs from the statistical mechanical point of view, a snapshot of the current status of research, as well as an educational presentation of the different ways to look at the same problem.

11.2 KEMs with No Saving: A Micro-canonical Ensemble Approach (Exponential Distribution)

In the basic versions of KEMs, N agents exchange a quantity x which represents the wealth. The state of the system is characterized by the set of variables $\{x_i\}$, $(i = 1, 2, \ldots, N)$. The total wealth is conserved (here set conventionally equal to one),

$$X = x_1 + x_2 + \cdots + x_{N-1} + x_N = 1, \tag{11.1}$$

The evolution of the system is carried out according to a prescription, which defines the trading rule between agents. Dragulescu and Yakovenko introduced the following simple model (Dragulescu and Yakovenko 2000): at every time step two agents i and j, with wealths x_i and x_j respectively, are extracted randomly and a random redistribution of the sum of the wealths of the two agents takes place, according to

$$\begin{aligned} x_i' &= \varepsilon(x_i + x_j), \\ x_j' &= (1 - \varepsilon)(x_i + x_j), \end{aligned} \tag{11.2}$$

where ε is a uniform random number $\varepsilon \in (0, 1)$, while x_i' and x_j' are the agent wealths after the "transaction". This rule is equivalent to a random reshuffling of the total wealth of the two interacting agents. After a large number of iterations, the system relaxes toward an equilibrium state characterized by a wealth distribution $f(x)$ which numerical experiments show to be perfectly fitted by an exponential function,

$$f(x) = \frac{1}{\langle x \rangle} \exp(-x/\langle x \rangle)\,, \tag{11.3}$$

where $\langle x \rangle$ is the average wealth of the system. The exponential function (see below for a demonstration) represents the distribution of kinetic energy in a two-dimensional gas, $D = 2$, with an effective temperature T defined by $T = 2\langle x \rangle / D \equiv \langle x \rangle$. This result can be demonstrated analytically using different methods, such as the Boltzmann equation, entropy maximization, etc., as discussed in the following sections. Here we start with a geometrical derivation of the exponential distribution based on the micro-canonical hypothesis.

Since the total amount of wealth $X = \sum_i x_i$ is conserved, the system is isolated and its state evolves on the positive part of the hyperplane defined by Eq. (11.1) in the configuration space. The surface area $S_N(X)$ of an equilateral N-hyperplane of side X is given by

$$S_N(X) = \frac{\sqrt{N}}{(N-1)!} X^{N-1}. \tag{11.4}$$

If the ergodic hypothesis is assumed, each point on the N-hyperplane is equiprobable. The probability density $f(x_i)$ of finding agent i with value x_i is proportional to the $(N-1)$- dimensional area formed by all the points on the N-hyperplane having the ith coordinate equal to x_i. If the ith agent has coordinate x_i, the $N-1$ remaining agents share the wealth $X - x_i$ on the $(N-1)$-hyperplane defined by

$$x_1 + x_2 \cdots + x_{i-1} + x_{i+1} \cdots + x_N = X - x_i, \tag{11.5}$$

whose surface area is $S_{N-1}(X - x_i)$. Defining the coordinate θ_N as

$$\sin\theta_N = \sqrt{\frac{N-1}{N}}, \tag{11.6}$$

then it can be shown that

$$S_N(X) = \int_0^N S_{N-1}(X - x_i) \frac{dx_i}{\sin\theta_N}. \tag{11.7}$$

Hence, the surface area of the N-hyperplane for which the ith coordinate is between x_i and $x_i + dx_i$ is proportional to $S_{N-1}(X - x_i)dx_i/\sin\theta_N$. Taking into account the normalization condition, one obtains

$$f(x_i) = \frac{1}{S_N(E)} \frac{S_{N-1}(E - x_i)}{\sin\theta_N} = (N-1)E^{-1}\left(1 - \frac{x_i}{E}\right)^{N-2} \to \frac{1}{\langle x \rangle} \exp\left(-x_i/\langle x \rangle\right), \tag{11.8}$$

where the last term was obtained in the limit of large N introducing the mean wealth per agent $\langle x \rangle = X/N$. From a rigorous point of view the Boltzmann factor $\exp(-x_i/\langle x \rangle)$ is recovered only in the limit $N \gg 1$ but in practice it is a good approximation also for small values of N. This exponential distribution has been shown to agree well with real data in the intermediate wealth range (Dragulescu and Yakovenko 2001a, b) but in general it does not fit real distributions, neither at very low nor at very high values of wealth. Thus, some improvements of this minimal model are required.

11.3 KEMs with Saving: The Maxwell Velocity Distribution Approach (Γ-Distribution)

We now turn to a more general version of kinetic exchange model. In this model, a saving propensity parameter λ, with $0 \leq \lambda < 1$, is assigned to agents, representing the minimum fraction of wealth saved during a trade. Models of this type have also been proposed in the social science by Angle in the 80s (Angle 1983, 1986, 1993, 2002) and were rediscovered as an extension of the model considered above in Chakraborti and Chakrabarti (2000), Chakraborti (2002), Chakraborti and Patriarca (2008). As a working example, for clarity here we consider a simple model (Chakraborti and Chakrabarti 2000) in which the evolution law is defined by the trading rule

$$\begin{aligned} x_i' &= \lambda x_i + \varepsilon(1-\lambda)(x_i + x_j)\,, \\ x_j' &= \lambda x_j + \bar{\varepsilon}(1-\lambda)(x_i + x_j)\,, \end{aligned} \tag{11.9}$$

where ε and $\bar{\varepsilon} = 1 - \varepsilon$ are two random numbers from a uniform distribution in (0, 1). In this model, while the wealth is still conserved during each trade, $x_i' + x_j' = x_i + x_j$, only a fraction $(1 - \lambda)$ of the wealth of the two agents is reshuffled between them during the trade. The system now relaxes toward an equilibrium state in which the exponential distribution is replaced by a Γ-distribution (Abramowitz and Stegun 1970)—or at least it is well fitted by it (this was also noted in Angle 1986). The Γ-distribution $\gamma_{\alpha,\theta}(\xi)$ has two parameters, a scale-parameter θ and a shape-parameter α, and it can be written as

$$\gamma_{\alpha,\theta}(x) = \frac{1}{\theta\Gamma(\alpha)}\left(\frac{x}{\theta}\right)^{\alpha-1}\exp(-x/\theta)\,, \tag{11.10}$$

where $\Gamma(\alpha)$ is the Γ-function. Notice that the Γ-distribution only depends on the ratio x/θ; namely, $\theta\gamma_{\alpha,\theta}(x)$ is a dimensionless function of the rescaled variable $\xi = x/\theta$. In the numerical simulations of the model, in which one assigns the initial average wealth $\langle x\rangle$ which is constant in time, the equilibrium distribution $f(x)$ is just the Γ-distribution with the λ-dependent parameters

$$\alpha(\lambda) = 1 + \frac{3\lambda}{1-\lambda} = \frac{1+2\lambda}{1-\lambda}\,, \tag{11.11}$$

$$\theta(\lambda) = \frac{\langle x\rangle}{\alpha} = \frac{1-\lambda}{1+2\lambda}\langle x\rangle\,. \tag{11.12}$$

As the saving propensity λ varies from $\lambda = 0$ toward $\lambda = 1$, the parameter α continuously assumes all the values between $\alpha = 1$ and $\alpha = \infty$. Notice that for $\lambda = 0$ the model and correspondingly the equilibrium distribution reduce to those of the model considered in the previous section.

At first sight it may seem that, in going from the exponential shape wealth distribution (obtained for $\lambda = 0$) to the Γ-distribution (corresponding to a $\lambda > 0$) the link between wealth-exchange models and kinetic theory is in a way lost, but, in fact, the Γ-distribution $\gamma_{\alpha,\theta}(x)/\theta$ represents just the canonical Boltzmann equilibrium distribution for a perfect gas in $D = 2\alpha$ dimensions and a temperature (in energy units) $\theta = k_B T$, where T is the absolute temperature. This can be easily shown—in the case of an (integer) number of dimensions D—also from the Maxwell velocity distribution of a gas in d dimensions. Setting in the following $\theta = k_B T$, the normalized Maxwell probability distribution of a gas in D dimensions is

$$f(v_1, \ldots, v_D) = \left(\frac{m}{2\pi\theta}\right)^{D/2} \exp\left(-\sum_{i=1}^{D} \frac{mv_i^2}{2\theta}\right), \tag{11.13}$$

where v_i is the velocity of the ith particle. The distribution (11.13) depends only on the velocity modulus v, defined by $mv^2/2 = \sum_{i=1}^{D} mv_i^2/2$, and one can then integrate the distribution over the $D - 1$ angular variables to obtain the velocity modulus distribution function $f(v)$. With the help of the expression for the surface $\sigma_D(r)$ of a hypersphere of radius r in D dimensions,

$$\sigma_D(r) \equiv \sigma_D^1 r^{D-1} = \frac{2\pi^{D/2}}{\Gamma(D/2)} r^{D-1}, \tag{11.14}$$

where σ_D^1 is the expression for a unit-radius sphere, one obtains

$$f(v) = \frac{2}{\Gamma(D/2)} \left(\frac{m}{2\theta}\right)^{D/2} v^{D-1} \exp\left(-\frac{mv^2}{2\theta}\right), \tag{11.15}$$

and then, by changing variable from the velocity v to the kinetic energy $x = mv^2/2$,

$$f(x) = \frac{1}{\Gamma(D/2)\theta} \left(\frac{x}{T}\right)^{D/2-1} \exp\left(-\frac{x}{T}\right), \tag{11.16}$$

which is just the distribution in Eq. (11.10) if one sets $\alpha = D/2$.

Notice that in order to construct a KEM with a given effective temperature θ and dimension D, it is not sufficient to fix the saving parameter λ. Following Eqs. (11.11)–(11.12), one has to assign both λ and the average energy $\langle x \rangle$. Using the relation $\alpha = D/2$, Eqs. (11.11)–(11.12) can be rewritten as

$$D(\lambda) = 2\left(1 + \frac{3\lambda}{1-\lambda}\right) = \frac{2(1+2\lambda)}{1-\lambda}, \tag{11.17}$$

$$\theta(\lambda) = \frac{2\langle x \rangle}{D}. \tag{11.18}$$

Inverting these equations, one has a simple recipe for finding the suitable values of λ and $\langle x \rangle$ that set the system dimension D and temperature θ to the required values, e.g. first fixing λ from D and then $\langle x \rangle$ using D and θ,

$$\lambda = \frac{D-2}{D+4}, \tag{11.19}$$

$$\langle x \rangle = \frac{D\theta}{2}. \tag{11.20}$$

Here the second equation can be recognized as an expression of the equipartition theorem.

It can be noticed that if λ is fixed (depending on the corresponding D) while the value of $\langle x \rangle$ is kept constant in all the simulations with different λ's, then from Eq. (11.18) a system with larger dimension D will have lower temperature θ (and therefore a shape of the probability distribution function with smaller width). From these equations one can also notice the existence of a minimum value for the system dimension, $D_{\text{min}} = 2$, corresponding to the minimum value $\lambda = 0$. As λ increases in the interval $\lambda \in (0, 1)$, D also increases monotonously diverging eventually for $\lambda \to 1$. This is the specific result of the model considered in this section and the minimum value D_{min} is different in different KEMs.

11.4 KEMs in *D* Dimensions: A Variational Approach

As an alternative, equivalent, and powerful approach, one can use the Boltzmann approach based on the minimization of the system entropy in order to obtain the equilibrium distribution (Chakraborti and Patriarca 2009). The method can provide both the exponential distribution as well as the Γ-distribution obtained in the framework of wealth-exchange models with a saving parameter $\lambda > 0$, a natural effective dimension $D > 2$ being associated to systems with $\lambda > 0$.

The representative system is assumed to have D degrees of freedom, $q_1, \ldots, q_D$ (e.g. the particle momenta in a gas), and a homogeneous quadratic Hamiltonian X,

$$X(q_1, \ldots, q_D) \equiv X(q^2) = \frac{1}{2}(q_1^2 + \cdots + q_D^2) = \frac{1}{2}q^2, \tag{11.21}$$

where $q = (q_1^2 + \cdots + q_D^2)^{1/2}$ is the distance from the origin in the D-dimensional q-space. As an example, the D coordinates q_i can represent suitably rescaled values of the velocities so that Eq. (11.21) provides the corresponding kinetic energy function. The expression of the Boltzmann entropy of a system described by D continuous variables $q_1, \ldots, q_D$, is

$$S_D[q_1, \ldots, q_D] = -\int dq_1 \ldots \int dq_D \, f_D(q_1, \ldots, q_D) \ln[f_D(q_1, \ldots, q_D)]. \tag{11.22}$$

The system is subjected to the constraints on the conservation of the total number of systems (i.e. normalizing to one for a probability distribution function) and of the total wealth (implying a constant average energy $\bar{x}$), expressed by

$$\int dq_1 \dots \int dq_D \, f_D(q_1, \dots, q_D) = 1 \,, \tag{11.23}$$

$$\int dq_1 \dots \int dq_D \, f_D(q_1, \dots, q_D) X(q_1, \dots, q_D) = \bar{x} \,, \tag{11.24}$$

They can be taken into account using the Lagrange method, i.e. by a variation (with respect to the distribution $f_D(q_1, \dots, q_D)$) of the functional

$$S_{\text{eff}}[f_D] = \int dq_1 \dots \int dq_D \, f_N(q_1, \dots, q_D)\{\ln[f_D(q_1, \dots, q_D)] + \mu + \beta X(q^2)\}, \tag{11.25}$$

where μ and β are two Lagrange multipliers. The invariance of the Hamiltonian, depending only on the modulus q, allows the transformation from Cartesian to polar coordinates. Integrating over the $(D-1)$ coordinates spanning the solid angle with the help of the expression (11.14) for the surface of the hyper sphere, one obtains

$$S_{\text{eff}}[f_1] = \int_0^{+\infty} dq \, f_1(q) \left[\ln \left(\frac{f_1(q)}{\sigma_D^1 \, q^{D-1}} \right) + \mu + \beta X(q) \right] \tag{11.26}$$

where the probability density $f_D(q_1, \dots, q_D)$ in the D-dimensional space was expressed with the reduced probability density $f_1(q)$ in the one-dimensional q-space,

$$f_1(q) = \sigma_D^1 \, q^{D-1} f_D(q) \,. \tag{11.27}$$

Finally, transforming from q to the energy variable $x = q^2/2$, one obtains the probability distribution function

$$f(x) = \frac{dq(x)}{dx} \, f_1(q)|_{q=q(x)} = \frac{f_1(q)|_{q=q(x)}}{\sqrt{2x}} \,, \tag{11.28}$$

where $q(x) = \sqrt{2x}$ from Eq. (11.21). In terms of the new variable x and distribution $f(x)$, from Eq. (11.26) one obtains the functional

$$S_{\text{eff}}[f] = \int_0^{+\infty} dx \, f(x) \left[\ln \left(\frac{f(x)}{\sigma_D^1 \, x^{D/2-1}} \right) + \mu + \beta x \right] , \tag{11.29}$$

Varying this functional with respect to $f(x)$, $\delta S_{\text{eff}}[f]/\delta f(x) = 0$, leads to the equilibrium Γ-distribution in Eq. (11.10) with rate parameter $\beta = 1/\theta$ and the same shape parameter $\alpha = D/2$.

11.5 Kinetic Theory Approach to a D-Dimensional Gas

The deep analogy between kinetic wealth-exchange models of closed economy systems, where agents exchange wealth at each trade, and kinetic gas models, in which energy exchanges take place at each particle collisions, was clearly noticed in Mandelbrot (1960). The analogy can be justified further by studying the microscopic dynamics of interacting particles in the framework of standard kinetic theory.

In one dimension, particles undergo head-on collisions, in which they can exchange the total amount of energy they have, i.e. a fraction $\omega = 1$ of it. Alternatively, one can say that the minimum fraction of energy that a particle saves in a collision is in this case $\lambda \equiv 1 - \omega = 0$. In the framework of wealth-exchange models, this case corresponds to the model of Dragulescu and Yakovenko mentioned above (Dragulescu and Yakovenko 2000), in which the *total* wealth of the two agents is reshuffled during a trade.

In an arbitrary (larger) number of dimensions, however, this does not take place, unless the two particles are travelling exactly along the same line in opposite verses. On average, only a fraction $\omega = (1 - \lambda) < 1$ of the total energy will be lost or gained by a particle during a collision, that is most of the collisions will be practically characterized by an energy saving parameter $\lambda > 0$. This corresponds to the model of Chakraborti and Chakrabarti (2000), in which there is a fixed maximum fraction $(1 - \lambda) > 0$ of wealth which can be reshuffled.

Consider a collision between two particles in an N-dimensional space, with initial velocities represented by the vectors $\mathbf{v}_{(1)} = (v_{(1)1}, \ldots, v_{(1)N})$ and $\mathbf{v}_{(2)} = (v_{(2)1}, \ldots, v_{(2)N})$. For the sake of simplicity, the masses of the all the particles are assumed to be equal to each other and will be set equal to 1, so that momentum conservation implies that

$$\begin{aligned} \mathbf{v}'_{(1)} &= \mathbf{v}_{(1)} + \Delta\mathbf{v} \,, \\ \mathbf{v}'_{(2)} &= \mathbf{v}_{(2)} - \Delta\mathbf{v} \,, \end{aligned} \tag{11.30}$$

where $\mathbf{v}'_{(1)}$ and $\mathbf{v}'_{(2)}$ are the velocities after the collisions and $\Delta\mathbf{v}$ is the momentum transferred. Conservation of energy implies that $\mathbf{v}'^2_{(1)} + \mathbf{v}'^2_{(2)} = \mathbf{v}^2_{(1)} + \mathbf{v}^2_{(2)}$ which, by using Eq. (11.30), leads to

$$\Delta\mathbf{v}^2 + (\mathbf{v}_{(1)} - \mathbf{v}_{(2)}) \cdot \Delta\mathbf{v} = 0 \,. \tag{11.31}$$

Introducing the cosines r_i of the angles α_i between the momentum transferred $\Delta\mathbf{v}$ and the initial velocity $\mathbf{v}_{(i)}$ of the i th particle ($i = 1, 2$),

$$r_i = \cos\alpha_i = \frac{\mathbf{v}_{(i)} \cdot \Delta\mathbf{v}}{v_{(i)}\, \Delta v} \,, \tag{11.32}$$

where $v_{(i)} = |\mathbf{v}_{(i)}|$ and $\Delta v = |\Delta\mathbf{v}|$, and using Eq. (11.31), one obtains that the modulus of momentum transferred is

$$\Delta v = -r_1 v_{(1)} + r_2 v_{(2)} \,. \tag{11.33}$$

From this expression one can now compute explicitly the differences in particle energies x_i due to a collision, that are the quantities $x'_i - x_i \equiv (\mathbf{v}'^2_{(i)} - \mathbf{v}^2_{(i)})/2$. With the help of the relation (11.31) one obtains

$$\begin{aligned} x'_1 &= x_1 + r_2^2 x_2 - r_1^2 x_1 \,, \\ x'_2 &= x_2 - r_2^2 x_2 + r_1^2 x_1 \,. \end{aligned} \tag{11.34}$$

The equivalence to KEMS should now appear clearly. First, the number r_i's are squared cosines and therefore they are in the interval $r \in (0, 1)$. Furthermore, they define the initial directions of the two particles entering the collision, so that they can be considered as random variables if the hypothesis of molecular chaos is assumed. In this way, they are completely analogous to the random coefficients $\varepsilon(1-\lambda)$ [or $(1-\varepsilon)(1-\lambda)$] appearing in the formulation of KEMs, with the difference that they cannot assume all values in (0, 1), but are limited in the interval $(0, 1-\lambda)$. However, in general the r_i^2's are not uniformly distributed in (0, 1) and the most probable values $\langle r_i^2 \rangle$ drastically depend on the space dimension, which is at the base of their effective equivalence with the KEMs: the greater the dimension D, the smaller the $\langle r_i^2 \rangle$, since the more unlikely it becomes that the corresponding values $\langle r_i \rangle$ assume values close to 1 and the more probable that instead they assume a small value close to $1/D$. This can be seen by computing their average—over the incoming directions of the two particles or, equivalently, on the orientation of the initial velocity $\mathbf{v}_{(i)}$ of one of the two particles and of the momentum transferred $\Delta \mathbf{v}$, which is of the order of $1/D$.

The $1/D$ dependence of $\langle r_i^2 \rangle$ well compares with the wealth-exchange model with $\lambda > 0$, in which a similar relation is found between the average value of the corresponding coefficients $\varepsilon(1-\lambda)$ and $\bar{\varepsilon}(1-\lambda)$ in the evolution equations (11.9) for the wealth exchange and the effective dimensions $D(\lambda)$, Eq. (11.17): since ε is a uniform random number in (0, 1), then $\langle \varepsilon \rangle = 1/2$ and inverting $D = D(\lambda)$, Eq. (11.17), one finds $\langle (1-\varepsilon)(1-\lambda) \rangle = \langle \varepsilon(1-\lambda) \rangle = (1-\lambda)/2 = 3/(D+4)$.

11.6 Numerical Simulations of a D-Dimensional Gas

Besides the various analytical considerations discussed above, the close analogy with kinetic theory allows one to resort to molecular dynamics simulations also to study a D-dimensional system. It is instructive to obtain the very same Boltzmann energy distributions discussed above from molecular dynamics (MD) simulations. For the sake of simplicity we consider the distribution of kinetic energy of a gas, since it is known that at equilibrium it relaxes to the Boltzmann distribution with the proper number D of dimensions of the gas *independently of the inter-particle potential*.

For clarity we start with the case $D = 2$. In fact, as discussed above when considering the model defined in Eq. (11.9), the case of the minimum dimension $D = 2$

is characterized by an equilibrium distribution which is a perfect exponential. We have performed some numerical simulation of a Lennard-Jones gas in $D = 2$ dimensions using the leapfrog algorithm (Frenkel and Smit 1996), with a small system of $N = 20$ particles in a square box of rescaled size $L = 10$, for a simulation time $t_{\rm tot} = 10^4$, using an integration time step $\delta t = 10^{-4}$ and averaging the energy distribution over 10^5 snapshots equidistant in time. Reflecting boundary conditions were used and a "repulsive Lennard-Jones" $U(r)$ interaction potential between particles was assumed,

$$U(r) = \varepsilon\left[(R/r)^6 - 1\right]^2 \quad \text{for} \quad r < R\,, \tag{11.35}$$

$$= 0\,, \qquad \text{for} \quad r \geq R\,, \tag{11.36}$$

representing a purely repulsive potential decreasing monotonously as the interparticle distance r increases, as far as $R = 1$, where the potential becomes (and remains) zero for all larger values of r.

As examples, the results of the kinetic energy distribution in $D = 2$ dimensions are shown in Fig. 11.1. The corresponding results for a gas in a cubic box in $D = 1$ and $D = 3$ dimensions, with the same parameters are shown in Figs. 11.2 and 11.3. Notice that in all figures the "MD" curve represents the result of the molecular simulation, while the "D = ..." curve is the corresponding Γ-distribution with shape parameter $\alpha = D/2$ and scale parameter $\theta = T = 1$.

In $D = 1$ dimensions, Newtonian dynamics predicts that the velocity distribution does not change with time in a homogeneous gas, since at each collision the two

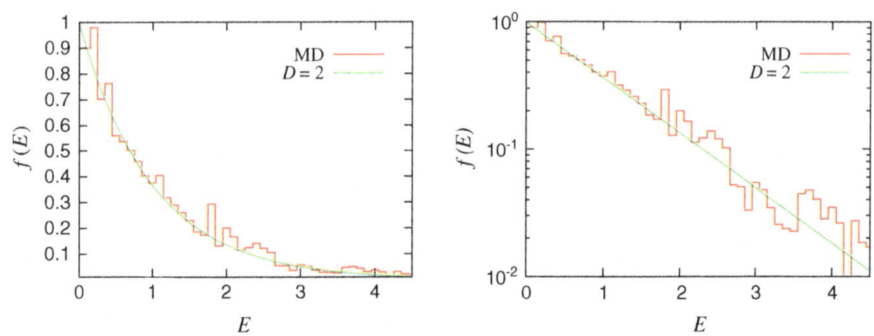

Fig. 11.1 Example of kinetic energy distribution for a gas in $D = 2$ dimensions in the linear (*left*) and semilog (*right*) scale. In the particular case of $D = 2$, the distribution is a perfect exponential

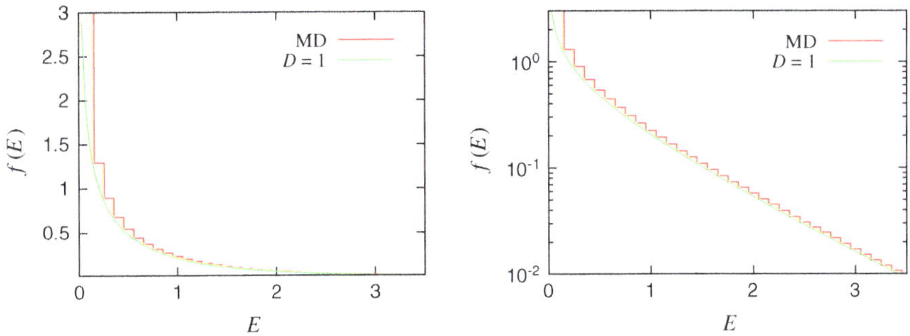

Fig. 11.2 Example of kinetic energy distribution for a gas in $D = 1$ dimensions in the linear (*left*) and semilog (*right*) scale

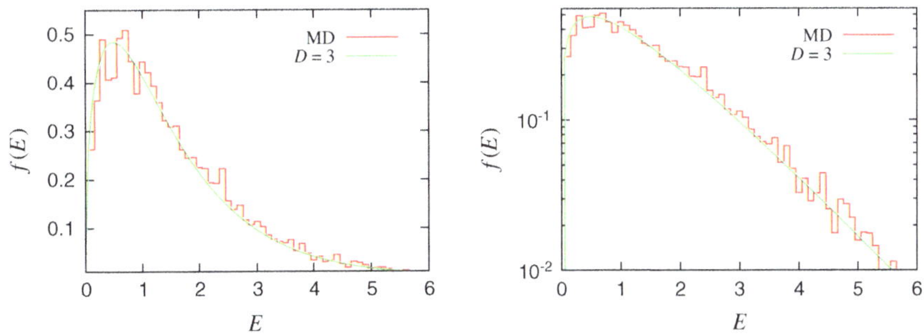

Fig. 11.3 Example of kinetic energy distribution for a gas in $D = 3$ dimensions in the linear (*left*) and semilog (*right*) scale. In the case of $D = 3$ dimensions, the Boltzmann distribution is proportional to $\sqrt{x}\,\exp(-\beta x)$

colliding particles simply exchange their momenta. Therefore, only in the case $D = 1$, we have added a Langevin thermostat at $T = 1$ (damping coefficient $\gamma = 0.5$) in order to induce a thermalization of the system.

11.7 Conclusions

KEMs represent one more approach to the study of prototypical statistical systems in which N units exchange energy and for this reason they certainly have a relevant educational dimension (Patriarca and Chakraborti 2013). This dimension is emphasized in this contribution by presenting different approaches to the same model and to obtain the corresponding canonical Boltzmann distribution in D dimensions.

There are other reasons for their relevance and for the interest they have attracted:

(a) KEMs have the peculiarity that the system dimension D can be easily tuned continuously. Letting it assume real values by changing the parameters regulating the energy exchanges, makes them interesting to study various other statistical systems.
(b) KEMs have by now been used in interdisciplinary physics in various topics such as modeling of wealth exchange and opinion dynamics;
(c) but they also appear in condensed matter problems such as fragmentation dynamics.

Acknowledgements M.P. and E.H. acknowledge support from the Institutional Research Funding IUT (IUT39-1) of the Estonian Ministry of Education and Research. A.S. is grateful to Council of Scientific and Industrial Research (CSIR), New Delhi, India for the financial support. A.C. acknowledges financial support from grant number BT/BI/03/004/2003(C) of Government of India, Ministry of Science and Technology, Department of Biotechnology, Bioinformatics Division and University of Potential Excellence-II grant (Project ID-47) of the Jawaharlal Nehru University, New Delhi, India.

References

M. Abramowitz and I. A. Stegun, editors. *Handbook of Mathematical Functions*. Dover, N.Y., 1970.

J. Angle. The surplus theory of social stratification and the size distribution of personal wealth. In *Proceedings of the American Social Statistical Association, Social Statistics Section*, pages 395–400, Alexandria, VA, 1983.

J. Angle. The surplus theory of social stratification and the size distribution of personal wealth. *Social Forces*, 65:293–326, 1986.

J. Angle. Deriving the size distribution of personal wealth from *The rich get richer, the poor get poorer*. *J. Math. Sociol.*, 18:27, 1993.

J. Angle. The statistical signature of pervasive competition on wage and salary incomes. *J. Math. Sociol.*, 26:217–270, 2002.

A. Chakraborti. Distribution of money in model markets of economy. *Int. J. Mod. Phys. C*, 13(10):1315–1321, 2002.

A. Chakraborti and B. K. Chakrabarti. Statistical mechanics of money: How saving propensity affects its distribution. *Eur. Phys. J. B*, 17:167–170, 2000.

A. Chakraborti and M. Patriarca. Gamma-distribution and Wealth Inequality. *Pramana*, 71:233–243, 2008.

A. Chakraborti and M. Patriarca. Variational Principle for the Pareto Power Law. *Physical Review Letters*, 103:228701, 2009.

A. Dragulescu and V. M. Yakovenko. Statistical mechanics of money. *Eur. Phys. J. B*, 17:723–729, 2000.

A. Dragulescu and V. M. Yakovenko. Evidence for the exponential distribution of income in the USA. *Eur. Phys. J. B*, 20:585, 2001a.

A. Dragulescu and V. M. Yakovenko. Exponential and power-law probability distributions of wealth and income in the United Kingdom and the United States. *Physica A*, 299:213, 2001b.

D. Frenkel and B. Smit. *Understanding Molecular Simulations: from algorithms to applications*. Academic Press, 1996.

B. Mandelbrot. The Pareto-Levy law and the distribution of income. *Int. Econ. Rev.*, 1:79, 1960.

M. Patriarca and A. Chakraborti. Kinetic exchange models: From molecular physics to social science. *Am. J. Phys.*, 81(8):618–623, 2013.

Chapter 12
The Microscopic Origin of the Pareto Law and Other Power-Law Distributions

Marco Patriarca, Els Heinsalu, Anirban Chakraborti and Kimmo Kaski

Abstract Many complex systems are characterized by power-law distributions, beginning with the first historical example of the Pareto law for the wealth distribution in economic systems. In the case of the Pareto law and other instances of power-law distributions, the power-law tail can be explained in the framework of canonical statistical mechanics as a statistical mixture of canonical equilibrium probability densities of heterogeneous subsystems at equilibrium. In this picture, each subsystem interacts (weakly) with the others and is characterized at equilibrium by a canonical distribution, but the distribution associated to the whole set of interacting subsystems can in principle be very different. This phenomenon, which is an example of the possible constructive role of the interplay between heterogeneity and noise, was observed in numerical experiments of Kinetic Exchange Models and presented in the conference "*Econophys-Kolkata-I*", hold in Kolkata in 2005. The 2015 edition, taking place ten years later and coinciding with the twentieth anniversary of the 1995 conference hold in Kolkata where the term "Econophysics" was introduced, represents an opportunity for an overview in a historical perspective of this mechanism within the framework of heterogeneous kinetic exchange models (see also *Kinetic exchange models as D-dimensional systems* in this volume).

M. Patriarca (✉) · E. Heinsalu
NICPB–National Institute of Chemical Physics and Biophysics,
Rävala 10, 10143 Tallinn, Estonia
e-mail: marco.patriarca@kbfi.ee

E. Heinsalu
e-mail: els.heinsalu@kbfi.ee

A. Chakraborti
SCIS–School of Computational & Integrative Sciences,
Jawaharlal Nehru University, New Delhi 110067, India
e-mail: anirban@jnu.ac.in

K. Kaski
Department of Computer Science, Aalto University School of Science,
P.O.Box 15500, FI-00076 Aalto, Finland
e-mail: kimmo.kaski@aalto.fi

F. Abergel et al. (eds.), *Econophysics and Sociophysics: Recent Progress and Future Directions*, New Economic Windows,
DOI 10.1007/978-3-319-47705-3_12

We also propose a generalized framework, in which both quenched heterogeneity and time dependent parameters can comply constructively leading the system toward a more robust and extended power-law distribution.

12.1 Introduction

The scaling properties of the power-law tails encountered in the distributions associated to many complex systems are a signature of some underlying processes of self-organization. A goal of Complex Systems Theory is to understand what are these processes and how they produce the observed distributions. The first example of power-law distribution, found in a field far from the targets of traditional physics, is related to the Pareto power-law of wealth distribution (Chatterjee et al. 2005).

Currently, the origin of power-law tails and their microscopic interpretation are still open problems, as shown by various proposals of mechanisms responsible for their appearance. For instance, Tsallis has suggested an extensive generalization of the Boltzmann entropy (Tsallis 1988), leading to a power-law probability distribution, while an alternative form of the Gibbs distribution was suggested in Treumann and Jaroschek (2008).

The first goal of this contribution is to provide a detailed discussion of how power-law distributions can be explained as due to the *diversity* of the components of the system under study within the framework of canonical statistical mechanics. In such systems the equilibrium distribution $f(x)$ of the relevant variable x is the statistical mixture of the different equilibrium distributions of the heterogeneous subsystems, each one with the shape of a Boltzmann-Gibbs-type canonical equilibrium distributions. A power-law tail can appear in the distribution $f(x)$ as the outcome of the superposition of the heterogeneous distributions of the subsystems. The general mechanism is formalized in Sect. 12.2. This mechanism was first suggested in the 2004 paper by Chatterjee, Chakrabarti, and Manna *Pareto law in a kinetic model of market with random saving propensity*" (Chatterjee et al. 2004), in which a power-law tail was first obtained from numerical experiments on heterogeneous Kinetic Exchange Models (KEMs). It was then described in detail through additional numerical experiments of KEMs and shown to represent a possible explanation of the Pareto power-law in economics by various groups (Bhattacharya et al. 2005; Chatterjee and Chakrabarti 2005; Patriarca et al. 2005) in the "*Econophys-Kolkata-I*" Conference, hold in Kolkata in 2005 [see Chatterjee et al. (2005) for the full list of contributions]. In the tenth anniversary of that conference and in the twentieth anniversary of the 1995 conference hold in Kolkata where the term "Econophysics" was introduced, the new 2015 edition represents an appropriate place for an overview of the topic. For this reason, KEMs are here reviewed from a historical perspective in Sect. 12.3—see also the contribution on *Kinetic exchange models as D-dimensional systems* in this volume. The same mechanism was later recognized as a possible general framework for describing power-laws as a collective effect taking place in *heterogeneous* complex systems not only of economical nature, see Patriarca et al.

(2016), but also in e.g. heterogeneous molecular assemblies, considered below in Sect. 12.5 with a simple exactly solvable kinetic model. The interpretation of power-law tails illustrated in this paper as a *diversity-induced phenomenon* may clarify and unify different instances of power-law tails appearing in the statistical distributions of many complex systems.

Furthermore, we discuss and merge in a generalized framework the diversity-based mechanism and the *superstatistics*, suggested in Beck and Cohen (2003) to describe the appearance of power-law distributions within non-equilibrium statistical mechanics.

12.2 A General Formulation of the Mechanism of Diversity-Induced Power-Laws

The mechanism producing a power-law distribution starting from the heterogeneous character of the units composing the system can be given a very simple and general probabilistic formulation. We consider a system $\mathscr{S}$ composed of N heterogeneous units, with K ($K \leq N$) different types of units, in which each unit n ($n = 1, \ldots, N$) can be of one type k among the K possible types. The system can then be partitioned in K homogeneous subsystems, $\mathscr{S} = \cup_{k=1}^{K} \mathscr{S}_k$, where $\mathscr{S}_k \cap \mathscr{S}_{k'} = 0$ if $k \neq k'$, by assigning each unit n to the corresponding subsystem $\mathscr{S}_k$ depending on its type k. One can introduce a statistical weight $p_k = N_k/N$ measuring the relative size of each subsystem $\mathscr{S}_k$, where N_k is the number of units of type k in the global system. For clarity the partition of the system $\mathscr{S}$ is kept fixed, i.e., the populations N_k of the homogeneous subsystems are constant in time.

Units are assumed to be described by some quantity (e.g. energy or wealth) measured by the variable x. The corresponding (global) probability distribution function $f(x)$ that the variable of a subsystem assumes the value x can be operatively constructed by measuring the frequency of occurrence of the value x of a randomly chosen subsystem in the limit of a large number of measurements. The *partial* probability densities $f_k(x)$ that the variable of a system of a given type k has the value x, can be operatively constructed in a similar way, recording in each measurement the value of x *and* the type of the unit k. Units of the same subsystem are assumed to follow the same dynamics and relax toward the same equilibrium distribution.

The relation between the global distribution $f(x)$ and the partial distributions $f_i(x)$ is given in probability theory by the *law of total probability* (Feller 1966), which provides an expression for the global distribution $f(x)$ as a *statistical mixture*—i.e. as a weighted sum—of the partial distributions $f_i(x)$,

$$f(x) = \sum_k f_k(x)\, p_k \equiv \sum_k P(x|k)P(k)\,, \tag{12.1}$$

where the last equality reminds that p_k coincides with $P(k)$, the probability to extract a unit of the kth type (independently of its variable value) belonging to the subsystem $\mathscr{S}_k$, while $f_k(x)$ coincides with the conditional probability $P(x|k)$ that, if the unit extracted is of the kth type, the variable has the value x. In Eq. (12.1) both the global probability distribution $f(x)$ and each partial probability density $f_k(x)$ are assumed to be normalized, i.e., $\int dx f(x) = 1$ and $\int dx f_k(x) = 1$, implying that also the statistical weights are normalized according to $\sum_k p_k = 1$.

The set of weights $\{p_k\}$ characterizes the level and type of heterogeneity of the system $\mathscr{S}$. Therefore Eq. (12.1) expresses the global probability distribution $f(x)$ directly in terms of the *diversity* of the system, defined by $\{p_k\}$. Notice that in general the global equilibrium distribution $f(x)$ of a heterogeneous composite system may have a very different shape with respect to those of the subsystems and that no prediction about $f(x)$ can be done without detailed information about the subsystems $\mathscr{S}_k$. Therefore it is possible that the statistical mixture in Eq. (12.1) will produce a distribution with a power-law tail even if none of the $f_k(x)$ has a power-law tail, depending on the details of the system considered, namely on (a) the form of the equilibrium partial distributions $f_k(x)$ and (b) the heterogeneity of the system as defined by the statistical weights p_k's.

Here below we illustrate some examples of heterogeneous systems presenting an equilibrium distribution with a diversity-induced power-law distribution. The cases of exactly solvable models are particularly instructive in that the shapes of the partial distributions $f_k(x)$ are known and an arbitrary weight distribution $\{p_k\}$ can be given as input parameter, while the other examples discussed are more phenomenological in nature but for this reason they are interesting from the point of view of complex systems theory.

12.3 An Introduction to KEMs

The Pareto law of wealth distribution is probably the first example of power-law distribution ever reported, see Fig. 12.1 for a real example. Even if an experimentally-based verification of its nature in terms of a statistical mixture of distributions of heterogeneous economic units—e.g. suitably defined categories of companies—as described in the previous section and as predicted in the framework of KEMs (see below) is still missing, there are various reasons to study KEMs in this respect.

First, KEMs have been proposed and rediscovered various times and in this way justified and motivated as basic models of wealth exchange using different approaches and from different points of view. This gives us enough confidence to state that at least the basic idea at the heart of KEMs (see below) must play a relevant role not only as a paradigm for the explanation of the appearance of power-law tails in general as a diversity-induced effect, which is a main topic discussed in the present paper, but also as an explanation for the specific and economically interesting case of the Pareto power-law characterizing wealth distributions.

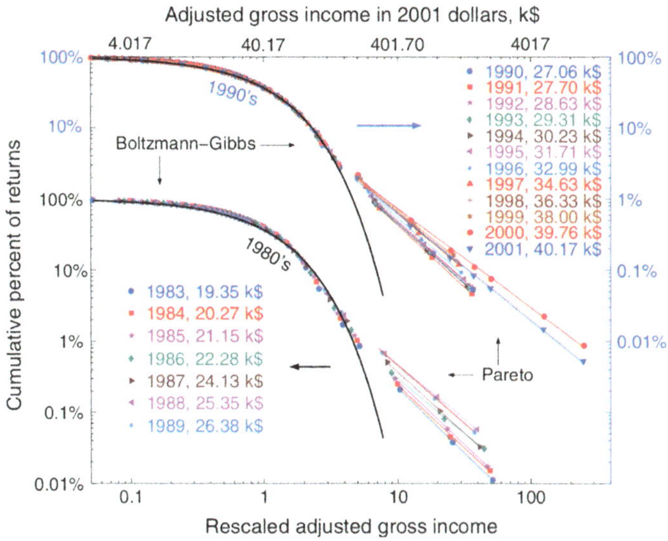

Fig. 12.1 Cumulative probability distribution of the net wealth, composed of assets (including cash, stocks, property, and household goods) and liabilities (including mortgages and other debts) in the United Kingdom shown on log-log (main panel) and log-linear (inset) scales. Points represent the data from the Inland Revenue, and solid lines are fits to the Boltzmann-Gibbs (exponential) and Pareto (power) distributions (Silva and Yakovenko 2005)

Furthermore, KEMs deserve a detailed discussion for their relevance from the historical point of view. In fact, it was just 10 years ago, during the 2005 edition of this same Econophys Conference Series hold in Kolkata (Chatterjee et al. 2005) that some groups reported for the first time about the possibility that in a heterogeneous economic systems modeled according to a KEM the various subsystems—i.e. the economic agents of the model—could relax toward standard equilibrium states characterized by canonical equilibrium distributions of wealth, while the corresponding marginal wealth distribution can result in a particularly realistic shape of wealth distribution exhibiting both a power-law at large values of wealth (with a realistic value of the power exponent) and an exponential shape at intermediate values, in agreement with real data of wealth and money distributions (Bhattacharya et al. 2005; Chatterjee and Chakrabarti 2005; Patriarca et al. 2005). These contributions were in turn stimulated by previous papers by A. Chatterjee, B.K. Chakrabarti, and S.S. Manna, who showed for the first time how in a KEM with a diversified set of economic agents a power-law would replace the exponential shape of the Boltzmann-Gibbs distribution.

Finally, some versions of KEMs turn out to be exactly solvable while for others we know what is very probably the exact solution (despite it has not been shown yet rigorously to be such), thus making them a particularly clear and detailed example of power-law formation.

The models known today as KEMs represent simple archetypal models of statistical mechanics that have been re-appearing from time to time in different fields and problems starting from the first studies of probability theory and statistical mechanics—in fact they can be considered to be related to the urn models. Currently kinetic exchange models have been applied also to other problems in molecular theory and opinion dynamics.

The specific versions of kinetic exchange models considered in the present paper were introduced with a precise economical problem in mind, namely that of describing and predicting the shape of the wealth distributions observed. The first quantitative modeling in this direction was put forward in the framework of social sciences more than 30 years ago in Angle (1983, 1986). The models of Angle were inspired by the Surplus Theory in Economics and introduced important novelties in the modeling of wealth exchanges, such as a pair-wise exchange dynamics with random fluctuations. The shape of the final equilibrium distribution obtained by Angle was surprisingly close to the Γ-distributions found in the analysis of real data. A related model was introduced in finance in Bennati (1988a, b, 1993), and was studied numerically through Monte Carlo simulations. Also that model leads to an equilibrium wealth distribution coinciding with the Boltzmann distribution (the Γ-distribution can be considered as a particular case of Boltzmann distribution).

In the physics community—more precisely in the framework of the field now known as *Econophysics*—different versions of kinetic exchange models were introduced, by S. Ispolatov, P.L. Krapivsky, and S. Redner (1998), by A. Chakraborti and B.K. Chakrabarti (2000), and by A. Dragulescu and V. Yakovenko (2000). In particular, the latter two papers were developed along a close and intriguing analogy between KEMs of economical systems and the physics of molecular fluids that stimulated in turn a long series of related works—KEMs had finally translated into a quantitative model an analogy already noticed many years before in Mandelbrot (1960). In principle the same effect could have been produced by Bennati (1988a, b, 1993) but unfortunately the papers were published in journals unaccessible to the physics community. Among the results obtained in the works which followed, it is of particular interest here that related to the explanation of the Pareto power-law as an "overlap of exponentials" (Patriarca et al. 2005) eventually formalized as a general mechanism of diversity-induced formation of power-laws (Patriarca et al. 2016) in terms of a *statistical mixture* of canonical equilibrium distributions. The concept of statistical mixture is well known in probability theory and in the present case is directly linked to the heterogeneous character of the economic agents—see previous section.

Further recent developments described below show that KEMs are a still active and stimulating research field, which will certainly provide new insights in many different disciplines (Patriarca and Chakraborti 2013).

12.3.1 The Homogeneous KEMs

We introduce the general structure of a KEM by a simple example. It is assumed that the N (minimally) interacting units $\{i\}$, with $i = 1, 2, \ldots, N$, are molecules of a gas with no interaction energy and the variables $\{x_i\}$ represent their kinetic energies, such that $x_i \geq 0$. The time evolution of the system proceeds by a discrete stochastic dynamics. A series of updates of the kinetic energies $x_i(t)$ are made at the discrete times $t = 0, 1, \ldots$. Each update takes into account the effect of a collision between two molecules. The time step, which can be set to $\Delta t = 1$ without loss of generality, represents the average time interval between two consecutive molecular collisions; that is, on average, after each time step Δt, two molecules i and j undergo a scattering process and an update of their kinetic energies x_i and x_j is made. The evolution of the system is accomplished as follows at each time t:

1. Randomly choose a pair of molecules i, j, with kinetic energies x_i, x_j, respectively; they represent the molecules undergoing a collision.
2. Compute the amount Δx_{ij} of kinetic energy exchanged, from the initial kinetic energies x_i, x_j and model parameters.
3. Perform the energy exchange between i and j by updating their kinetic energies,

$$x_i \to x_i - \Delta x_{ij}, \quad x_j \to x_j + \Delta x_{ij}\,, \tag{12.2}$$

   (the total kinetic energy is conserved during an interaction).
4. Set $t \to t + 1$ and go to step 1.

The form of the function Δx_{ij} depends on the specific model. Kinetic exchange models describe the dynamics at a microscopic level, based on single molecular collisions. Such a representation can be optimal in terms of simplicity and computational efficiency when the focus is on the energy dynamics, because particles are described by their energy degree of freedom w only, rather than by the entire set of their $2D$ position and momentum coordinates, for a D-dimensional system (Fig. 12.2).

As a first simple example, consider the reshuffling rule

$$x_i \to \varepsilon(x_i + x_j), \tag{12.3}$$

$$x_j \to (1 - \varepsilon)(x_i + x_j), \tag{12.4}$$

where ε is a stochastic variable drawn as a uniform random number between 0 and 1. This rule corresponds to a $\Delta x_{ij} = (1 - \varepsilon)x_i - \varepsilon x_j$ in Eq. (12.2). In this case, the algorithm we have outlined leads from arbitrary initial conditions to the Boltzmann-Gibbs energy distribution at equilibrium $f(x) = \beta \exp(-\beta x)$, where $\beta = 1/\langle x \rangle$ and $\langle x \rangle$ represents the mean energy of a single molecule. The theoretical derivations of this result using the Boltzmann transport equation, or entropy maximization principle, or simple probabilistic arguments, can be found in standard textbooks of statistical mechanics.

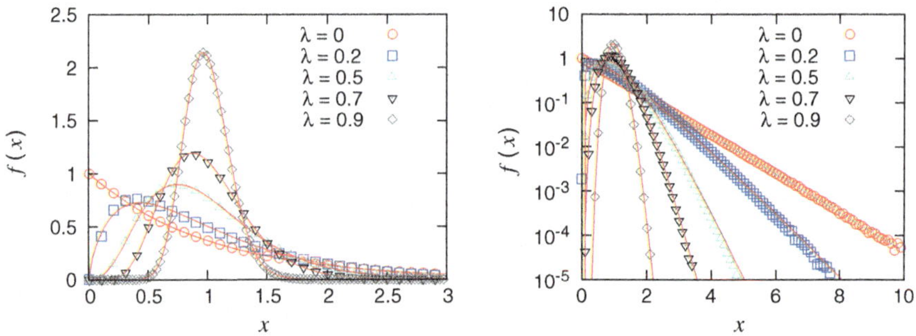

Fig. 12.2 Equilibrium wealth distributions in linear (*left*) and semi-log (*right*) scale for different values of the saving parameter λ and corresponding Γ-distribution fitting functions

As a more general example, consider the relaxation in energy space of a gas in D-dimensions. We assume that $D > 1$ because the momentum and energy distributions of a one-dimensional gas (where only head-on collisions occur) do not change with time. For a gas in D dimensions Δx_{ij} can be derived from energy and momentum conservation during a collision between particles i and j. If the respective D-dimensional vectors of the particle initial momenta are $\mathbf{p}_i$ and $\mathbf{p}_j$, we find (Chakraborti and Patriarca 2008)

$$\Delta x_{ij} = r_i x_i - r_j x_j \tag{12.5}$$

$$r_k = \cos^2 \alpha_k \quad (k = i, j) \tag{12.6}$$

$$\cos \alpha_k = \frac{\mathbf{p}_k \cdot \Delta \mathbf{p}_{ij}}{|\mathbf{p}_k| \, |\Delta \mathbf{p}_{ij}|}, \tag{12.7}$$

where $\cos \alpha_k$ is the direction cosine of momentum $\mathbf{p}_k$ $(k = i, j)$ with respect to the direction of the transferred momentum $\Delta \mathbf{p}_{ij} = \mathbf{p}_i - \mathbf{p}_j$. The numbers r_k can be assumed to be random variables in the hypothesis of molecular chaos.

We can now study the time evolution by randomly choosing at each time step two new values for r_k in Eq. (12.5) instead of maintaining a list of momentum coordinates, as is done in a molecular dynamics simulation. Then we use Eq. (12.2) to compute the new particle energies x_i and x_j. Note that the r_k's are not uniformly distributed in (0, 1), and thus some care has to be used in choosing the form of their probability distribution function. In fact, their distribution strongly depends on the spatial dimension D, their average value being $\langle r_k \rangle = 1/D$ (see Chakraborti and Patriarca (2008) for further details). The dependence of $\langle r_k \rangle$ on D can be understood from kinetic theory: the greater the value of D, the more unlikely it becomes that r_k assumes values close to $r_k = 1$ (corresponding to a one-dimensional-like head on collision). A possibility is to choose a uniform random distribution $f(r_k)$ limited in the interval $(0, r_{\max})$, with a suitable value of $r_{\max}$ yielding the same average value $1/D$, as in the model with a finite saving propensity illustrated below.

Simulations of this model system using random numbers in place of the r_k's in Eq. (12.5), for $D = 2$, give the equilibrium Boltzmann-Gibbs distribution: $f(x) = \beta \exp(-\beta x)$, where $\beta = 1/\langle x \rangle$. For $D > 2$, we obtain the D-dimensional generalization of the standard Boltzmann distribution (Patriarca et al. 2004a, b; Chakraborti and Patriarca 2008), namely the Γ- distribution (Abramowitz and Stegun 1970; Weisstein 2016) characterized by a shape parameter α equal to half spatial dimension,

$$f(w, \alpha, \theta) = \frac{w^{\alpha-1} e^{-w/\theta}}{\theta^{\alpha} \Gamma(\alpha)} \tag{12.8}$$

$$\alpha = D/2 \tag{12.9}$$

$$\theta = \langle w \rangle / \alpha \, . \tag{12.10}$$

The scale parameter θ of the Γ-distribution is fixed, by definition, by Eq. (12.10) (Abramowitz and Stegun 1970; Weisstein 2016). From the equipartition theorem in classical statistical mechanics, $w = D k_{\rm B} T/2$. Hence, we see that Eq. (12.10) identifies the scale parameter θ as the absolute temperature (in energy units) given by $\theta \equiv k_{\rm B}T = 1/\beta$. Therefore, the same Boltzmann factor, $\exp(-w/\theta)$, is present in the equilibrium distribution independently of the dimension D, and the prefactor $w^{\alpha-1}$ depends on D, because it takes into account the phase-space volume proportional to $p^d \propto w^{d/2}$, where p is the momentum modulus. In KEMs, one finds a relation between the effective dimension $D(\lambda)$ and the "saving parameter" λ, with $0 \leq \lambda \leq 1$. In general, the larger Δ, the closer to 1 is λ. In the case of some particular variants of KEMs it has been finally demonstrated rigorously in Katriel (2015) that the equilibrium distribution is a Γ-function. By inverting $\Delta(\lambda)$, one obtains that the average fraction of wealth exchanged during a trade is $1-\lambda \propto 1/D$ for $D \gg 1$, similarly to the energy exchanges during molecular collisions in a D-dimensional gas, where two molecules exchange on average a fraction of energy inversely proportional to the space dimension (Chakraborti and Patriarca 2008).

12.3.2 *The Heterogeneous KEMs*

An interesting generalization of the homogeneous kinetic exchange models discussed so far is the introduction of heterogeneity. Probably the most relevant applications of heterogeneous kinetic exchange models in the social sciences is the prediction of a realistic shape for the wealth distribution, including the Pareto power-law at the largest wealth values, compare Figs. 12.1 and 12.4. At a general level, heterogeneous KEMs are composed of agents with different saving parameters λ_i and have interesting physics analogues of dimensionally heterogeneous systems. For instance, in the case of a uniform distribution for the saving parameters, $\phi(\lambda) = 1$ if $\lambda \in (0, 1)$ and $\phi(\lambda) = 0$ otherwise, setting $n = D/2$, the dimension density has a power-law $\sim 1/n^2$, $P(n) = \phi(\lambda) d\lambda/dn = 3/(n+2)^2$ $(n \geq 1)$.

Considering again the model put forward in Chakraborti and Chakrabarti (2000), heterogeneity is introduced by diversifying the saving parameter λ, meaning that each λx_i is to be replaced by the corresponding term $\lambda_i x_i$, thus obtaining an exchanged wealth

$$\Delta x_{ij} = (1 - \varepsilon)(1 - \lambda_i)x_i - \varepsilon(1 - \lambda_j)x_j \,. \tag{12.11}$$

As a simple example, one can consider a set of heterogeneous agents with parameters λ_k uniformly distributed in the interval (0, 1). By repeating the simulations using Eq. (12.11), it is found that the shape of the separate equilibrium wealth distributions $f_k(x)$ of each agent k still retains a Γ-distribution form. However, the wealth distribution of the system $f(x)$, given by the sum of the wealth distributions of the single agents, $f(x) = \sum_i f_i(x)$, has an exponential form until intermediate x-values while a Pareto power-law develops at the largest values of x, see Fig. 12.3. Such a shape is in fact prototypical of real wealth distributions, compare Fig. 12.1. This shape of the equilibrium wealth distribution $f(x)$ is robust with respect to the details of the system and the other parameters, as long as the values of the λ_k are sufficiently spread over the whole interval $\lambda = (0, 1)$. In fact, it is the group of agents with $\lambda \approx 1$ that are crucial for the appearance of a power-law. This is well illustrated by the fact that a similar distribution shape is obtained from a quite different set of λ-parameters, namely from an agent population in which 99 % have a homogeneous $\lambda = 0.2$, while only 1 % of the population has a saving propensity spread in $\lambda = (0, 1)$, see Fig. 12.4. The way in which the single Γ-distributions of the subsystems comply to generate a power-law distribution is illustrated in Fig. 12.3, taken from Patriarca et al. (2005, 2006). The heterogeneous model necessarily uses a finite upper cutoff $\lambda_{\max} < 1$, when considering the saving parameter distribution, which directly determines the cutoff $x_{\max}$ of the wealth distribution, analogous to the cutoff observed in real

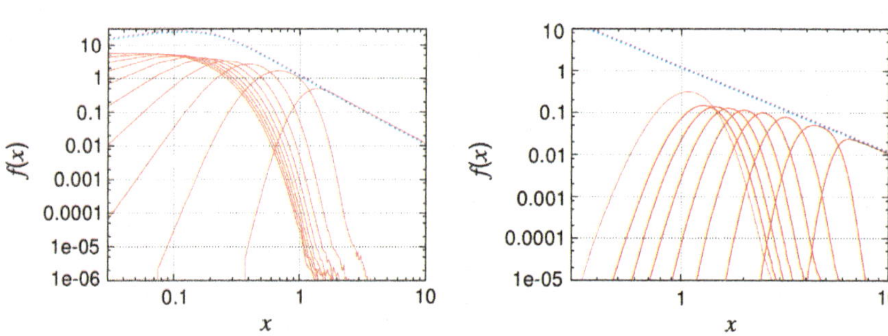

Fig. 12.3 Wealth distribution $f(x)$, from Patriarca et al. (2006), for uniformly distributed λ_k in the interval (0,1); $f(x)$ is here resolved into partial distributions $f_i(x)$, where each $f_i(x)$ is obtained counting the statistics of those agents with parameter λ_i in a specific sub-interval. *Left*: Resolution of $f(x)$ into ten partial distributions in the ten λ-subintervals (0, 0.1), (0.1, 0.2) …(0.9, 1). *Right*: The last distribution of the left figure in the λ-interval (0.9, 1) is in turn resolved into partial distributions obtained counting the statistics of agents with λ-subintervals (0.9, 0.91), (0.91, 0.92) …(0.99, 1). Notice how the power-law appears as a consequence of the superposition of the partial distributions

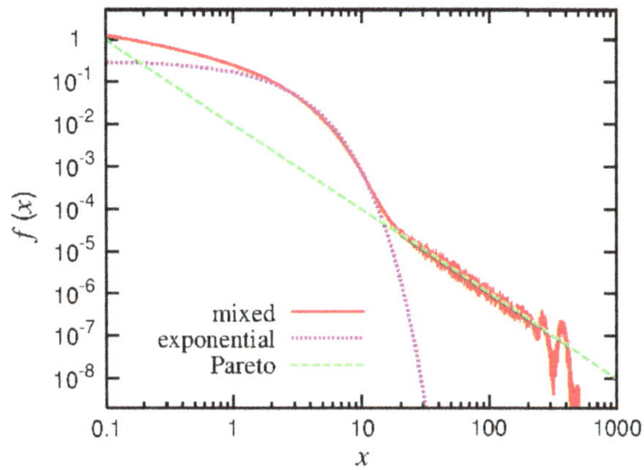

Fig. 12.4 Example of a realistic wealth distribution, from Patriarca et al. (2006). *Red* (*continuous*) *curve*: Wealth distribution obtained by numerical simulation of an agent population with 1 % of uniformly distributed saving propensities $\lambda_i \in (0, 1)$ and the rest of homogeneous saving propensities $\lambda_i = 0.2$. *Magenta* (*dotted*) *curve*: Exponential wealth distribution with the same average wealth, plotted for comparison with the distribution in the intermediate-wealth region. *Green* (*dashed*) *curve*: Power-law $\propto x^{-2}$ plotted for comparison with the large-income part of the distribution

distributions: the closer $\lambda_{\max}$ is to one, the larger $x_{\max}$ and the wider the interval in which the power-law is observed (Patriarca et al. 2006).

Also, the λ-cutoff is closely related to the relaxation process, whose time scales for a single agent i is proportional to $1/(1 - \lambda_i)$ (Patriarca et al. 2007). Thus, the slowest convergence rate is determined by $1 - \lambda_{\max}$. The finite λ-cutoff used in simulations of heterogeneous kinetic exchange models is not a limitation of the model, but reflects an important feature of real wealth distributions.

We notice that a derivation of the dynamics of the kinetic exchange models from microeconomics theory was proposed in Chakrabarti and Chakrabarti (2009), in terms of the *utility maximization principle* used in standard economic theory. The picture described here is instead that of agents as particles exchanging "money" in the place of energy in conserving two-body scattering, as in *entropy maximization* based on the kinetic theory of gases (Chakraborti and Patriarca 2009).

12.4 Power-Laws in Complex Networks

A type of power-law distribution which has not been mentioned so far is that associated to an underlying complex topology.

As a first example we compare a free diffusion process on a homogeneous network and that on a scale-free network. We first consider a homogeneous lattice of M sites, in which each site i ($i = 1, \ldots, M$) is connected to the same number $k_i = k$ of first neighbors. Such a lattice is an example of a dimensionally homogeneous network providing a discrete representation of a D-dimensional space. In the case of the square lattice structure, the dimension D is related to the degree k as $D = k/2$. An unbiased uniform diffusion process of X walkers hopping between the M sites of the lattice relaxes toward a uniform load distribution $f(x) = \text{const}$, with the same average load at each node i given by $x_i = X/M$.

On the other hand, a heterogeneous network with degree distribution $g(k)$ cannot be given a straightforward geometrical interpretation and in general represents a space with a highly complex topology. In particular, no unique dimension can be assigned, so that it can be regarded as a dimensionally heterogeneous space. One can estimate a local dimensionality from the connectivity, in analogy with the homogeneous square lattice, by introducing for each node i the local dimension $D_i = k_i/2$. At equilibrium, free diffusion of X walkers on such a network produces a stationary state with an average load x_i proportional to the degree, $x_i = \bar{x}k_i$, where the average flux per link and direction $\bar{x}$ is fixed by normalization, $\bar{x} = X/K$, with $X = \sum_i x_i$ and $K = \sum_j k_j$. It follows from probability conservation that the load distribution at equilibrium $f(x)$ is directly determined by the degree distribution $g(x)$,

$$f(x) = g(k)dk/dx = g(x/\bar{x})/\bar{x}\,. \tag{12.12}$$

In the important case of a scale-free network with $g(k \gg 1) \sim 1/k^p$, one has a power-law tail in the load distribution, $f(x) \sim 1/x^p$, with the same exponent p. A close relation between degree distribution and equilibrium density, analogous to Eq. (12.12) valid for the case of free diffusion, can be expected for any quantity x diffusing through a network. For instance, in the case of the Zipf law, such a relations is known to hold, if written language is regarded as a random walk across the complex network with nodes given by words and links between words which are neighbors in a text.

12.5 A Heterogeneous Polymeric Fluid

As an example of a standard system presenting a diversity-induced power-law distribution, we consider a theoretical model made up of an assembly of harmonic polymers. This is a simple and exactly solvable model, yet it is general in the sense that the inter-particle harmonic potentials can be thought to describe the small displacements of the normal modes with respect to the equilibrium configuration of a more general nonlinear system. We assume that polymers consist of different numbers of monomers, i.e., they have different numbers of degrees of freedom, see Fig. 12.5, and study the potential energy distribution (similar considerations hold also for the distribution of kinetic energy or velocity). Notice that such a model can also be used to study a general system composed of subsystems with different dimensions or numbers of degrees of freedom. The hypothesis of non-interacting polymers is made, in the same spirit of the statistical mechanical treatment of a perfect gas, even if a weak interaction is understood to be present in order to bring the system toward thermal equilibrium, implying that each polymer undergoes independent statistical

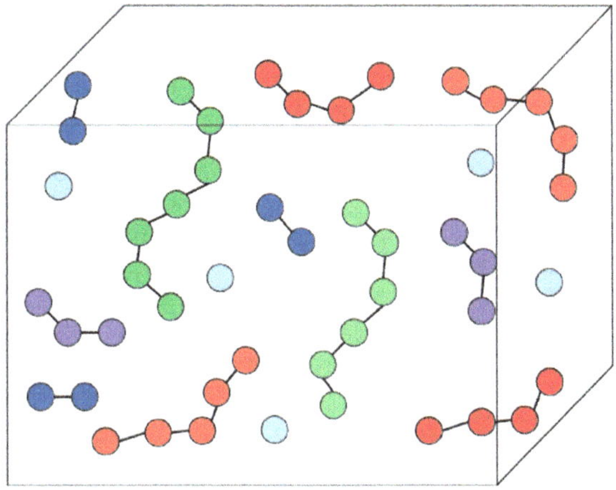

Fig. 12.5 A prototypical model of system presenting a diversity-induced power-law tail in the (kinetic as well as potential) energy distribution is an assembly of harmonic polymers with different numbers of monomers, see text for details

fluctuations. It is convenient to start from the homogeneous system, composed of identical subsystems with D harmonic degrees of freedom. Using suitably rescaled coordinates $\mathbf{q} = \{q_i\} = \{q_1, q_2, \ldots, q_D\}$, the energy function can be written in the form $x(\mathbf{q}) = (q_1^2 + \cdots + q_D^2)/2$. The equilibrium energy distribution coincides with the standard Gibbs energy distribution of a D-dimensional harmonic oscillator. After integrating out the angular variables in the space $\mathbf{q}$, it reduces to a Γ-function of order $n = D/2$ (Patriarca et al. 2004b),

$$f_n(x) = \beta\, \gamma_n(\beta x) \equiv \frac{\beta}{\Gamma(n)} (\beta x)^{n-1} \exp(-\beta x)\,, \quad n = D/2\,. \tag{12.13}$$

Here β is the inverse temperature. The same result is obtained through a variational principle from the Boltzmann entropy, see the contribution on the KEMs in D dimensions in this volume. The result presented there for the entropy $S_n[f_n]$ and the use of the method of the Lagrange multipliers can be directly generalized for the analogous problem of a heterogeneous system with different dimensions, i.e.,

$$S[\{f_n\}] = \int dn\, P(n) \int_0^{+\infty} dx\, f_n(x) \left\{ \ln\left[\frac{f_n(x)}{\sigma_{2n}\, x^{n-1}} \right] + \mu_n + \beta x \right\}, \tag{12.14}$$

where the fractions $P(n)$ of units with dimension $D = 2n$ have been introduced, with $\sum_n P(n) = 1$ and Γ_{2n} is the hypersurface in Δ dimensions. Different Lagrange multipliers μ_n have been used since the fractions $P(n)$ are conserved separately, while a single temperature parameter β means that only the total energy is conserved.

The resulting average energy is $\langle x \rangle \langle D \rangle / 2\beta$, where $\langle D \rangle = 2\langle n \rangle = 2 \int dnP(n)n$ is the average dimension. The probability of measuring a value x of energy (independently of the unit type) is a statistical mixture (Feller 1966),

$$f(x) = \int dn\, P(n) f_n(x) = \int dn\, \frac{P(n)\beta}{\Gamma(n)} (\beta x)^{n-1} \exp(-\beta x). \tag{12.15}$$

While the distributions $f_n(x)$ have exponential tails, the asymptotic shape of the function $f(x)$ can be in general very different. It is possible to show that

$$f(x \gg \beta^{-1}) \approx \beta P(\beta x)\,, \tag{12.16}$$

if $P(n)$ decreases fast enough with increasing n. Thus, if $P(n)$ has a power-law tail in n then $f(x)$ has a power-law tail in x with the same exponent. Some examples are shown in Fig. 12.6, taken from Chakraborti and Patriarca (2009), to which the reader is referred for a detailed discussion. This result can be obtained considering values $\beta x \gg 1$ in Eq. (12.15), since the main contributions to the integral come from $n \approx \beta x \gg 1$ ($\gamma_n(\beta x)$ has its maximum at $x \approx n/\beta$ and $\gamma_n(\beta x) \to 0$ for small as well as larger x). Introducing the variable $m = n - 1$, Eq. (12.15) can be rewritten as

$$f(x) = \beta \exp(-\beta x) \int dm\, \exp[-\phi(m)]\,, \tag{12.17}$$

$$\phi(m) = -\ln[P(m+1)] - m \ln(\beta x) + \ln[\,\Gamma(m+1)]\,. \tag{12.18}$$

This integral can be estimated through the saddle-point approximation expanding $\phi(m)$ to the second order in $\varepsilon = m - m_0$, where $m_0 = m_0(x)$ locates the maximum of $\phi(m)$, defined by $\phi'(m_0) = 0$ and $\phi''(m_0) > 0$, and integrating:

$$f(x) \approx \beta \sqrt{2\pi / \phi''(m_0)}\, \exp[-\beta x - \phi(m_0)]\,. \tag{12.19}$$

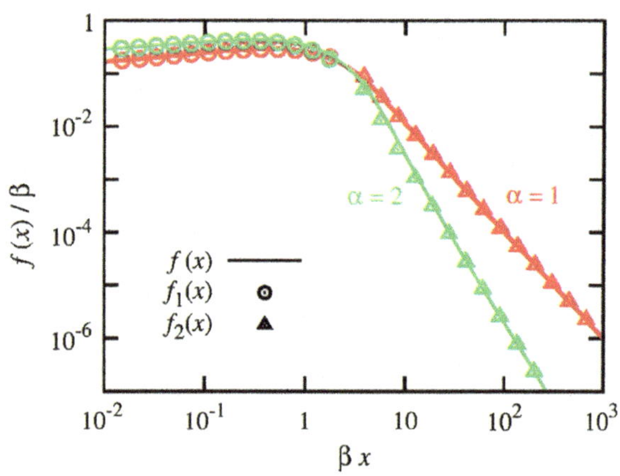

Fig. 12.6 Distribution $f(x)$ in Eq. (12.15) with $P(n) = \alpha / n^{1+\alpha}$ ($n \geq 1$), $P(n) = 0$ otherwise, for $\alpha = 1$ (*red*), $\alpha = 2$ (*green*). *Continuous lines*: Numerical integration of Eq. (12.15). *Triangles*: Saddle point approximation. *Circles*: Small-x limit. See text for details

Using the Stirling approximation in Eq. (12.18) one finds

$$\phi(m) \approx -\ln[P(m+1)] - m\ln(\beta x) + \ln(\sqrt{2\pi}) + (m+1/2)\ln(m) - m, \quad (12.20)$$

$$\phi'(n) \approx -P'(m+1)/P(m+1) - \ln(\beta x) + 1/2m + \ln(m), \quad (12.21)$$

$$\phi''(n) \approx P'^2(m+1)/P^2(m+1) - P''(m+1)/P(m+1) - 1/2m^2 + 1/m. \quad (12.22)$$

For general shapes of $P(n)$ which decrease fast enough one can neglect the terms containing P respect to $1/m$ as well as P'/P and $1/m$ respect to $\ln(m)$. Then the approximate solution of $\phi'(m_0) = 0$ is $m_0(x) \approx \beta x$ and using Eqs. (12.20)–(12.22) in Eq. (12.19) one has

$$f(x \gg \beta^{-1}) \equiv f_2(x) = \beta P(1 + \beta x)\,, \quad (12.23)$$

providing the asymptotic form of the density $f(x)$ in terms of the dimension density $P(n)$.

For the opposite limit of $f(x)$ at $x \ll \beta^{-1}$ one can set $\phi(n) \approx \phi(1) + \phi'(1)(n-1)$ in (12.17) and (12.18), to obtain

$$f(x \ll \beta^{-1}) \equiv f_1(x) = -[\beta P(1)\exp(-\beta x)]/[\ln(\beta x) + \gamma + P'(1)/P(1)], \quad (12.24)$$

where, from Eq. (12.18), we set $\phi(0) = \ln[P(1)]$, $\phi'(0) = -\gamma - \ln(\beta x) - P'(1)/P(1)$, with $\gamma = \psi(1) \equiv (d\ln[\Gamma(m)]/dm)_{m=1} \approx 0.57721$ being the Euler γ-constant.

In Fig. 12.6 the function $f_2(x)$ (triangles), given by Eq. (12.23), is compared at large x with the exact distribution $f(x)$ obtained by numerical integration of Eq. (12.15) (continuous lines) for the values $\alpha = 1, 2$ for the power-law density $P_\alpha(n)$. Also the corresponding density $f_1(x)$ (circles), given by Eq. (12.24), is shown at small βx.

12.6 A Generalized Framework for Power-Law Formation

As a concluding discussion and proposal for future research in this field, we compare on one side the diversity-related mechanism discussed above, describing power-law distributions as a heterogeneity-induced effect, and, on the other side, the mechanism referred to as *superstatistics*, introduced in the framework of non-equilibrium statistical mechanics (Beck and Cohen 2003; Beck 2006) to explain the appearance of power-laws as due to the long-time or large-space fluctuation of some parameter of the system, such as temperature: "*The basic idea underlying this approach is that there is an intensive parameter, for example the inverse temperature β or the energy dissipation in turbulent systems, that exhibits fluctuations on a large time scale (large as compared to internal relaxation times of the system under consideration)*" (Beck 2009).

The key mechanism that in superstatistics can lead to the appearance of a power-law is the interplay between the fast dynamics (describing e.g. the local Brownian motion) and the slow dynamics (associated e.g. to slow global variations of temperature). From the technical point of view, one can describe the effective probability distribution of the relevant variable as obtained through a *marginalization* procedure (Feller 1966) with respect to the possible values of the random parameter (from the physics point of view this means to integrate out the stochastic parameter) or as a *randomization* procedure (Feller 1966), i.e., an average of the probability distribution function of a (single) system over the values of some system parameter(s) varying slowly in time or space. In this way, without any further (exotic) assumptions beyond canonical statistical mechanics, superstatistics can lead to the appearance of a power-law tail in the distributions associated to different phenomena.

The two mentioned mechanisms are basically different in their physical interpretations and are in fact complementary to each other: in the diversity-induced power-law mechanism a quenched disorder is assumed and the system can in principle be in thermal equilibrium. Instead, superstatistics considers just the temperature (or the energy) fluctuations of the system as responsible for the power-law appearance.

However, both the diversity-based mechanism and superstatistics describe power-laws formation in terms of a superposition of otherwise canonical distribution, despite they are deeply different in their physical interpretation—marginal distributions due to the interplay of a set of heterogeneous constituent units in the first case and compound distributions resulting from random variation in time or space of some system parameters in the second case. At a methodological level, the justifications behind the two mechanisms share the intention to remain within the limits of canonical statistical mechanics. This suggests the interesting possibility that a generalized framework, in which the underlying ideas of the two methods are merged, i.e. taking into account both slow stochastic fluctuations of some parameters as well as internal heterogeneity, could provide a better description of fat-tailed distributions.

It is worth noting that such an approach was implicitly explored in Chatterjee et al. (2004) when a power-law tail was first observed in a numerical experiment with KEMs. In fact, besides diversifying the agents, the authors also introduced a random resetting at fixed periods of time of all the (different) saving parameters of the agents, which could be interpreted at a statistical level as a long time average over different values of the saving parameters, e.g. due to random fluctuations, in the spirit of superstatistics. As discussed already in the Kolkata conference of 2005 (Patriarca et al. 2005), the combination of these two different ways of averaging the wealth distribution over different values of the saving parameters turns out to be an effective procedure to extend the range of the power-law tail and to speed up its formation. More work is needed to clarify further the mutual advantage that diversity and long-time noise in the system parameters can have in producing fat-tailed distributions.

12.7 Conclusions

We have discussed the origin of power-law distributions in complex systems, i.e., how the heterogeneity of their constituent units lets a power-law tailed distribution emerge as a collective diversity-induced effect. A general formulation based on probability theory was given and a few examples of power-law tailed distributions were discussed in detail, together with the proposal of some further research. Much remains to be done for a deeper understanding of the results obtained so far and to fully explore their consequences.

Acknowledgements M.P. and E.H. acknowledge support from the Institutional Research Funding IUT(IUT39-1) of the Estonian Ministry of Education and Research. A.C. acknowledges financial support from grant number BT/BI/03/004/2003(C) of Government of India, Ministry of Science and Technology, Department of Biotechnology, Bioinformatics Division and University of Potential Excellence-II grant (Project ID-47) of the Jawaharlal Nehru University, New Delhi, India.

References

M. Abramowitz and I. A. Stegun, editors. *Handbook of Mathematical Functions*. Dover, N.Y., 1970.

J. Angle. The surplus theory of social stratification and the size distribution of personal wealth. In *Proceedings of the American Social Statistical Association, Social Statistics Section*, pages 395–400, Alexandria, VA, 1983.

J. Angle. The surplus theory of social stratification and the size distribution of personal wealth. *Social Forces*, 65:293–326, 1986.

C. Beck. Stretched exponentials from superstatistics. *Physica (Amsterdam)*, 365A:96, 2006.

C. Beck. Superstatistics in high-energy physics. *Eur. Phys. J. A*, 40:267–273, 2009.

C. Beck and E.G.D. Cohen. Superstatistics. *Physica A*, 322:267–275, 2003.

E. Bennati. *La simulazione statistica nell'analisi della distribuzione del reddito: modelli realistici e metodo di Monte Carlo*. ETS Editrice, Pisa, 1988a.

E. Bennati. Un metodo di simulazione statistica nell'analisi della distribuzione del reddito. *Rivista Internazionale di Scienze Economiche e Commerciali*, 35:735, August 1988b.

E. Bennati. Il metodo Monte Carlo nell'analisi economica. *Rassegna di lavori dell'ISCO*, X:31, 1993.

K. Bhattacharya, G. Mukherjee, and S. S. Manna. Detailed simulation results for some wealth distribution models in econophysics. In A. Chatterjee, S.Yarlagadda, and B. K. Chakrabarti, editors, *Econophysics of Wealth Distributions*, page 111. Springer, 2005.

Anindya S. Chakrabarti and Bikas K. Chakrabarti. Microeconomics of the ideal gas like market models. *Physica A*, 388:4151–4158, 2009.

A. Chakraborti and B. K. Chakrabarti. Statistical mechanics of money: How saving propensity affects its distribution. *Eur. Phys. J. B*, 17:167–170, 2000.

A. Chakraborti and M. Patriarca. Gamma-distribution and Wealth inequality. *Pramana J. Phys.*, 71:233, 2008.

A. Chakraborti and M. Patriarca. Variational principle for the Pareto power law. *Phys. Rev. Lett.*, 103:228701, 2009.

A. Chatterjee, B. K. Chakrabarti, and S. S. Manna. Pareto law in a kinetic model of market with random saving propensity. *Physica A*, 335:155, 2004.

A. Chatterjee and B.K. Chakrabarti. Ideal-Gas Like Markets: Effect of Savings. In A. Chatterjee, S. Yarlagadda, and B. K. Chakrabarti, editors, *Econophysics of Wealth Distributions*, pages 79–92. Springer, 2005.

A. Chatterjee, S. Yarlagadda, and B. K. Chakrabarti, editors. *Econophysics of Wealth Distributions - Econophys-Kolkata I*. Springer, 2005.

A. Dragulescu and V. M. Yakovenko. Statistical mechanics of money. *Eur. Phys. J. B*, 17:723–729, 2000.

W. Feller. volume 1 and 2. John Wiley & Sons, 2nd edition, 1966.

S. Ispolatov, P.L. Krapivsky, S. Redner, Wealth distributions in asset exchange models, *Eur. Phys. J. B*, 17:723–729, 1998

G. Katriel. Directed Random Market: the equilibrium distribution. *Eur. Phys. J. B*, 88:19, 2015.

B. Mandelbrot. The Pareto-Levy law and the distribution of income. *Int. Econ. Rev.*, 1:79, 1960.

M. Patriarca and A. Chakraborti. Kinetic exchange models: From molecular physics to social science. *Am. J. Phys.*, 81(8):618–623, 2013.

M. Patriarca, A. Chakraborti, and G. Germano. Influence of saving propensity on the power law tail of wealth distribution. *Physica A*, 369:723, 2006.

M. Patriarca, A. Chakraborti, E. Heinsalu, and G. Germano. Relaxation in statistical many-agent economy models. *Eur. J. Phys. B*, 57:219, 2007.

M. Patriarca, A. Chakraborti, and K. Kaski. Gibbs versus non-Gibbs distributions in money dynamics. *Physica A*, 340:334, 2004a.

M. Patriarca, A. Chakraborti, and K. Kaski. Statistical model with a standard gamma distribution. *Phys. Rev. E*, 70:016104, 2004b.

M. Patriarca, A. Chakraborti, K. Kaski, and G. Germano. Kinetic theory models for the distribution of wealth: Power law from overlap of exponentials. In A. Chatterjee, S. Yarlagadda, and B. K. Chakrabarti, editors, Econophysics of Wealth Distributions, page 93. Springer, 2005.

M. Patriarca, E. Heinsalu, L. Marzola, A. Chakraborti, and K. Kaski. *Power-Laws as Statistical Mixtures*, pages 271–282. Springer International Publishing, Switzerland, 2016.

A.C. Silva and V.M. Yakovenko. Temporal evolution of the thermal and superthermal income classes in the USA during 1983-2001. *Europhys. Lett.*, 69:304–310, 2005.

R. A. Treumann and C. H. Jaroschek. Gibbsian Theory of Power-Law Distributions. *Phys. Rev. Lett.*, 100:155005, 2008.

C. Tsallis. Possible generalization of Boltzmann-Gibbs statistics. *J. Stat. Phys.*, 52:479, 1988.

Eric W. Weisstein. Gamma Distribution 2016.

Chapter 13
The Many-Agent Limit of the Extreme Introvert-Extrovert Model

Deepak Dhar, Kevin E. Bassler and R.K.P. Zia

Abstract We consider a toy model of interacting extrovert and introvert agents introduced earlier by Liu et al. (Europhys. Lett. **100** (2012) 66007). The number of extroverts, and introverts is N each. At each time step, we select an agent at random, and allow her to modify her state. If an extrovert is selected, she adds a link at random to an unconnected introvert. If an introvert is selected, she removes one of her links. The set of N^2 links evolves in time, and may be considered as a set of Ising spins on an $N \times N$ square-grid with single-spin-flip dynamics. This dynamics satisfies detailed balance condition, and the probability of different spin configurations in the steady state can be determined exactly. The effective hamiltonian has long-range multi-spin couplings that depend on the row and column sums of spins. If the relative bias of choosing an extrovert over anF introvert is varied, this system undergoes a phase transition from a state with very few links to one in which most links are occupied. We show that the behavior of the system can be determined exactly in the limit of large N. The behavior of large fluctuations in the total number of links near the phase transition is determined. We also discuss two variations, called egalitarian and elitist agents, when the agents preferentially add or delete links to their least/most-connected neighbor. These shows interesting cooperative behavior.

D. Dhar (✉)
Tata Institute of Fundamental Research, Homi Bhabha Road, Mumbai, India
e-mail: ddhar@theory.tifr.res.in

K.E. Bassler
Department of Physics, University of Houston, Houston, TX 77204, USA
e-mail: bassler@uh.edu

R.K.P. Zia
Department of Physics, Virginia Polytechnic Institute and University, Blacksburg, VA 24061, USA
e-mail: rkpzia@vt.edu

F. Abergel et al. (eds.), *Econophysics and Sociophysics: Recent Progress and Future Directions*, New Economic Windows,
DOI 10.1007/978-3-319-47705-3_13

13.1 Introduction

In recent years, there has been a lot of interest in the study of networks. Many different types of networks have been studied: transportation networks like the railways or airlines networks (Sen et al. 2003), chemical reaction networks in a cell (Tyson et al. 2003), power grids etc. (Albert et al. 2004). A good review of the general theory may be found in Dorogovetsev et al. (2008), Barabasi (2009). In studies of networks in social sciences, some examples are the citation network of scientific publications (Chen and Redner 2010), small-world networks (Newman 2000). An important question that has attracted a lot of attention is the time evolution of social networks, in which the number of friends a particular agent interacts with evolves in time. In this context, a very interesting toy model was introduced by Liu et al. (2012), called the extreme introvert extrovert model. Numerical studies of this model revealed a phase transition, where the fractional number of links present undergoes a jump from a value near 0 to a value near 1, as a parameter in the model is varied continuously (Liu et al. 2013, 2014; Bassler et al. 2015). Recently, we have shown that this model can be solved exactly in the limit of large number of agents (Bassler et al. 2015). This article is a brief pedagogical account of these results. For details, the reader is referred to the original publication.

13.2 Definition of the Model

We consider a group consisting of N introvert and N extrovert agents. The agents can add or delete links connecting them to other agents. In so doing, in our model, they do not have to get permission from the agent to whom the link is being added or from whom the link is being removed. The introverts are assumed to be comfortable with only a few links, and extroverts like many. In a general model, the member of links an introvert likes to have is k_I, and an extrovert likes k_E, with $k_I < k_E$. We will consider the extreme case where $k_I = 0$, and $k_E = \infty$. Thus, and introvert does not like any links, and will delete a link, given an opportunity. On the other hand, an extrovert will try to add a link whenever possible. This model has been called the eXtreme Introvert-Extrovert model (XIE) (Liu et al. 2012, 2013, 2014).

A configuration of the system at any time is specified completely by an $N \times N$ matrix $\mathbb{A}$ whose entries are A_{ij}, with $A_{ij} = 1$, if there is link between the i-th introvert, and the j-th extrovert, and $A_{ij} = 0$ otherwise. The total number of configurations is 2^{N^2}. The model undergoes a discrete-time Markovian evolution defined by the following rules: At each time step, select an agent at random, and allow her to change the number of links connecting her to other agents. Any introvert has a probability $\frac{1}{(1+z)N}$, of being selected, and an extrovert has a probability $\frac{z}{(1+z)N}$ of being selected. Then, $z < 1$ corresponds to a bias favoring introverts, and $z > 1$ favors extroverts. If an introvert is selected, and has at least one link to an extrovert, she deletes one of the links connecting her to other agents at random. If she has no links, she does nothing, and the configuration remains unchanged. If an extrovert is selected, she will add a link to one of introverts not already linked to her. If she already has all links present, she does nothing.

13.3 Steady State of the XIE Model

We may think of the N^2 binary variables A_{ij} as Ising variables placed on an $N \times N$ square grid, and then the update rule corresponds to single-spin-flip dynamics of the model. Note that our rules have the symmetry of changing introverts to extroverts, and $A_{ij} \leftrightarrow 1 - A_{ij}$, $z \leftrightarrow 1/z$. This corresponds to the Ising model having spin-reversal symmetry with $z \leftrightarrow 1/z$.

In general, given some such set of update rules, it is very difficult to determine the probabilities of different configurations in the steady state exactly. The remarkable result about the XIE model is that in this case, the probabilities of transition between configurations $\mathscr{C}$ and $\mathscr{C}'$ satisfy the detailed balance condition, and one can write down the probability of different configurations in the steady state exactly. For the configuration $\mathscr{C}$, in which the i-th introvert has degree p_i, and the j-th extrovert has degree q_j, the steady-state probability Prob$^*(\mathscr{C})$ has a very pleasing form (Liu et al. 2012)

$$\mathscr{P}^*(\mathscr{C}) = \frac{1}{\Omega(z)} z^{\sum_i p_i} \prod_{i=1}^{N} (p_i!) \prod_{j=1}^{N} (N - q_j)! \tag{13.1}$$

Fwhere $\Omega(z)$ is a normalization constant.

We may define the negative logarithm of this probability as the ‘energy’ of the configuration $\mathscr{C}$, giving

$$H_{eff}(\mathscr{C}) = -\sum_{i=1}^{N} \log p_i! - \sum_{j=1}^{N} \log(N - q_j)! - \log(z) \sum_i p_i \tag{13.2}$$

We see that the effective hamiltonian has long-range couplings, and the energy of a configuration depends only on the row- and column- sums of the square array $\mathbb{A}$. Also, the energy function is non-extensive: the energy of the configuration with all links absent varies as $-N^2 \log N$. This non-extensivity causes no real problems, as all probabilities are well-defined, and averages of observables in steady state are well-behaved.

Monte Carlo simulations of the XIE model have shown that, for large N, the system seems to undergo a transition from a few-links phase for $z < 1$ to a phase in which almost all links present for $z > 1$. In the few-links phase, the average number of links per agent remains finite, of order 1, even as N tends to infinity, with fixed $z < 1$. Conversely, in the link-rich phase for $z > 1$, the average number of links per agent is nearly N, and the difference of this number from N remains finite, as N is increased to infinity.

The fact that energy depends only on the row- and column- sums, and thus only on $2N$ variables, instead of the N^2 variable A_{ij} explicitly, suggests that some kind of mean-field treatment may be exact for this problem. This turns out to be true, as we proceed to show, but the treatment needs some care, as the number of variables, and hence also their conjugate fields, tends to infinity, in the thermodynamic limit. If the energy of the system depended only on the total number of links in the system, a single mean-field variable conjugate to the net magnetization would have been sufficient.

13.4 Asymptotically Exact Perturbation Theory

We consider the low-density phase ($z < 1$) first. The case $z > 1$ is equivalent to this by the Ising symmetry discussed above. In the low density phase, the typical degree q_j of the j-th extrovert is much less than N. Then we have $(N - q_j)! \approx N!N^{-q_j}$. This suggests that we write for all $q \geq 0$

$$(N - q)!/N! = N^{-q} F(q, N) \tag{13.3}$$

with

$$F(q, N) = \prod_{r=1}^{q} \left(1 - \frac{r-1}{N}\right) \tag{13.4}$$

For $q \ll N$, $F(q, N)$ is nearly equal to 1. Then, since $\sum_j q_j = \sum_i p_i$, we can write the effective Hamiltonian for the random XIE model as

$$\mathscr{H}_{eff} = \mathscr{H}_0 + \mathscr{H}_{int} \tag{13.5}$$

where

$$\mathscr{H}_0 = -\sum_i \left[\ln (p_i!) + p_i \ln \frac{z}{N}\right] - N \ln(N!) \tag{13.6}$$

and

$$\mathscr{H}_{int} = -\sum_j \ln F(q_j, N) \tag{13.7}$$

If we ignore the effect of the "perturbation term" $\mathscr{H}_{eff}$, different introverts are independent, and one can sum over states of each introvert separately. This gives

$$\Omega_0 = (N!)^N [\omega_0]^N$$

with

$$\omega_0 = \sum_k z^k F(k, N_E)$$

For large N, F tends to 1, and we get

$$\omega_0 = 1 + z + z^2 + z^3 + ... = \frac{1}{1-z}, \tag{13.8}$$

which gives

$$\log \Omega(z) = N \ \log(N!) + N \ \log \ (\frac{1}{1-z}) + \dots . \tag{13.9}$$

In a systematic perturbation theory, we need to determine the behavior of the steady state of the system under $\mathscr{H}_0$. This is easily done. In particular, we can determine the degree distribution of introverts and extroverts. It is easily seen that for large N, the probability that an introvert has degree r has an exponential distribution: Prob (introvert has degree r) $= (1-z)z^r$. Here, the $p_i!$ factor in the weight of a configuration makes the usually expected Poisson form into a simple exponential. However, the degree distribution of the j-th extrovert is a sum of N mutually independent variables A_{ij}, hence it remains a Poisson distribution. Clearly the mean degree of extroverts is same as the mean degree of introverts, so the Poisson distribution has mean $\frac{z}{(1-z)}$, which determines it completely.

To lowest order in $(1/N)$, $\log F(q, N) = -q(q-1)/(2N)$. Thus, while the interaction hamilonian has different values for different configurations, and thus not a trivial c-number term, it is a sum of N different weakly correlated terms, and its mean is $\mathscr{O}(1)$, and fluctuations about the mean are smaller. It is easily seen that they are $\mathscr{O}(N^{-1/2})$, giving

$$\log \Omega(z) = N \ \log(N!) + N \ \log \ (\frac{1}{1-z}) + \mathscr{O}(1), \ \text{for } z < 1. \tag{13.10}$$

For $z > 1$, similar analysis, or the introvert-extrovert flip symmtery can be used to deduce that

$$\log \Omega(z) = N \ \log(N!) + N^2 \log z + N \ \log \ (\frac{1}{1-1/z}) + \mathscr{O}(1), \ \text{for } z > 1. \tag{13.11}$$

This is a remarkable result. Clearly, we get asymptotically exact result for $\log \Omega(z)$ up to the linear order in N using the hamiltonian $\mathscr{H}_0$. The effect of $\mathscr{H}_{int}$ is only a term of $\mathscr{O}(1)$ in $\Omega(z)$. In particular, in the large-N limit, the density of links is 0 for $z < 1$, and 1 for $z > 1$.

We note that these results are consistent with a scaling ansatz

$$\frac{[\Omega(z)]^{1/N}}{N!} = N^a f(\varepsilon N^b) \tag{13.12}$$

where $z = \exp(-\varepsilon)$, $a = b = 1/2$, and $f(0) =$ a finite constant, and $f(x)$ is continuous at $x = 0$. For large positive x, $f(x) \sim 1/x$. For $z > 1$, a similar form is obtained by the inversion symmetry, but the scaling functions for $\varepsilon > 0$ and $\varepsilon < 0$ are different, reflecting the 'first-order nature' of the transition.

13.5 Variants of XIE with Preferential Attachment

It is interesting to consider some variations on this general theme. We consider the case when the agent does not choose which link to add (or delete) at random, as done in the XIE model in Sect. 13.2, but decides on the basis of knowledge of degrees of the other nodes. We consider two variations.

Egalitarian agents: Here extrovert agents realize that the introverts regard links as burden, and attempts to distribute this burden as evenly as possible, and would add a link to the least connected introvert. Similarly, an introvert would cut a link to the *most* connected extrovert, as this action would make the other extroverts more equal.

Elitist agents: Here, we consider the opposite extreme. In this case, an extrovert prefers the most 'sociable' introvert, and adds a link to the *most* connected of the available introverts. Similarly, an introvert cuts a link to the *least* connected available extrovert.

These variations have a strong effect on the degree distribution in the steady state. Let us discuss egalitarians first. Then, at any time, the degree distribution of an agent will have only two possible values: k or $(k + 1)$, for some possibly time-dependent k. In the low density regime of this egalitarian XIE model, there are only a small number of contacts. It is easy to see that in the large-N limit, in the steady state, we have $k = 0$, and fractional number of introverts with exactly 1 contact is z. For the degree distribution of extroverts, it is easy to see that degree distribution is Poisson, with a mean value that increases with z. For $z > 1$, the only possible values of degree of an introvert are $N - 1$ and N.

The behavior of the degree distribution at the phase transition point $z = 1$ is particularly interesting. Here the fractional number of links can vary from 0 to 1, and the degree of an agent vary from 0 to N. However, the agents, by their cooperative behavior ensure that the inequality in the society always remains low. This is shown in Fig. 13.1, where we plot the time-dependent degree of two introverts, and two extroverts. While the actual degree varies in an unpredictable way, the four curves fall on top of each other, and are not distinguishable in the plot.

We now discuss the elitists case. To recall, here the extrovert agents prefer to link to one most sociable of the introverts (most connected), and introverts, to keep inversion symmetry, are assumed to delete their link to the least connected extrovert. This generates an interesting instability: Say, start with all links absent. Then, an extrovert will add a link to some randomly chosen introvert. Then all later extroverts selected at subsequent times will choose to link to the same agent. Her degree will tend to become N. Then the extroverts will choose a new "star" to link to. The degree

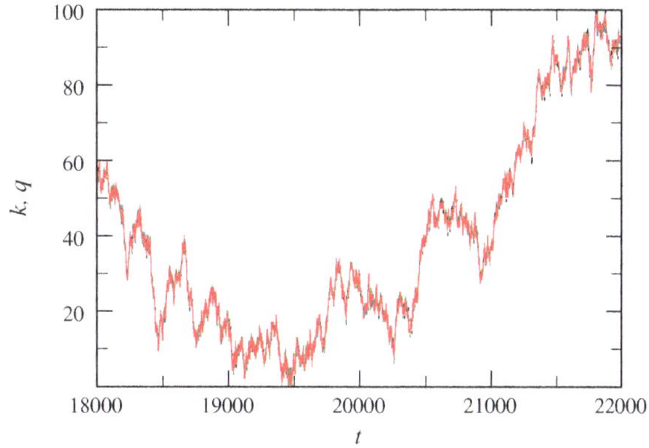

Fig. 13.1 Time trace of the degrees of two introvert k (*black*, *blue*), and of two extroverts, q (*red*, *green*) in a critical egalitarian XIE with $N = 100$. The graphs fall on top of each other, and are not distinguishable. Here the unit of t is a sweep. Taken from (Bassler et al. 2015)

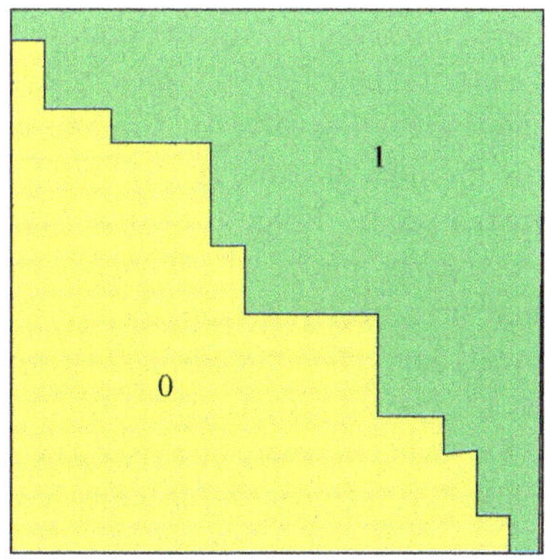

Fig. 13.2 The incidence matrix after reordering. The 1's and zeroes are separated by a single staircase-like interface. Here *green* = 1, *yellow* = 0

distribution of introverts becomes very unequal. The question what happens in the steady state?

Interestingly, it turns out that in this case, for all z, the degree distribution is rather wide. Let us try to understand this. Clearly, argument sketched above needs elaboration. When a 'star' introvert's degree becomes comparable to N, often an updating extrovert would be already connected to this introvert, and the time between changes of degree increases if the degree is closer to N. Meanwhile, other new stars are already in the making. This produces a broad distribution of degrees. If we inspect the incidence matrix $\mathbb{A}$ at different times during the evolution of the system, we do not see much pattern in it directly. However, if we look at the same matrix after permuting the rows and columns to matrix so that both introverts and extroverts are arranged according to their degrees in ascending order, we see much more structure. It is found, and easy to prove a posteriori, that in the sorted matrix, all 1's come in a single block with no holes, and similarly with zeroes (Fig. 13.2). There is a single staircase-shaped interface that separates 1's and zeros, and this fluctuates with time. Clearly, if we have this property at one time, it will be preserved by the dynamical rules.

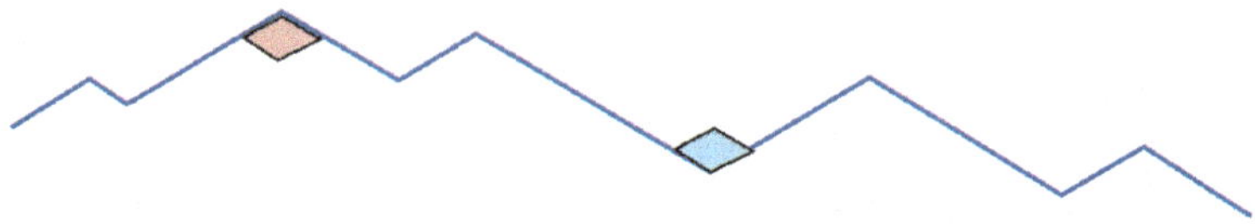

Fig. 13.3 Evolution rules for the interface: height can change at the maxima (*pink diamond*), or the minima (*blue diamond*), at a rateproportional to length of side on the *right*

The total number of accessible matrices in steady state is $\leq \binom{2N}{N}(N!)^2$, which is much less than the number of possible 2^{N^2}. Thus, the configurations are much simpler to describe, when the permutation symmetry between agents is factored out, and one works with the equivalence classes of configurations under the symmetry group. As time evolves, the interface describing the configuration will evolve (we have to reorder the agents according to their degree distribution at each time). One can write the evolution rules in terms of the interface model. It can be seen that the interface will evolve by flipping at corners: height can change at the maxima, or the minima, at a rate proportional to length of side on the right (Fig. 13.3). But this model seems difficult to solve exactly. On general grounds, since it is an interface fluctuation model where height can increase or decrease with equal probability, it may be expected to be in the universality class of Edwards-Wilkinson model, where the dynamical exponent is 2, i.e. relaxation time for an interface of length L varies as L^2. For the elitists case, this corresponds to relaxation time being approximately $\mathcal{O}(1)$ sweeps (where one sweep of the system is L^2 attempts which updates each link on the average once), for large N.

For different values of the z, the slope of the interface changes, but qualitative behavior of relaxation remains the same. Hence, we find that the elitists self-organize into a critical state, where the degree distribution of agents is wide, for all z.

13.6 Summary and Concluding Remarks

In this article, we discuss a simple toy model of a dynamical networks, where the state of different links keeps changing with time, but the system has a non-trivial steady state. As a function of the bias parameter z, the system undergoes a phase transition from a state with few links to a state with most links occupied. We showed that one can develop a perturbation theory, which becomes asymptotically exact for large N, at first order of perturbation. The corresponding state has non-trivial many-body correlations. We also discussed variations of the basic XIE model where the agents attach to one of the most- or least- connected agents, and showed that the steady state can have a wide degree distribution of degrees of agents. An interesting open problem is a theory to calculate the scaling function $f(x)$ in Eq. (13.12) exactly. It is hoped that future research will lead to a better understanding of this model.

Acknowledgements DD and RKPZ thank the Galileo Galilei Institute for Theoretical Physics for hospitality and the INFN for partial support during the summer of 2014. This research is supported in part by the Indian DST via grant DST/SR/S2/JCB-24/2005, and the US NSF via grants DMR-1206839 and DMR-1507371.

References

Albert R, Albert I and Nakarado G L, Phys. Rev. **E 69** (2004) 025103.
Barabasi A L, Science **325** (2009) 412.
Bassler K E, Dhar D and Zia R K P 2015 *J. Sta. Mech. Theory Exp.***2015** P07013.
K. E. Bassler, W. Liu, B. schmittmann and R. K. P. Zia, Phys. Rev. **E 91**, (2015) 042102.
Chen P and Redner S, J. Informetrics, **4** (2010) 278.
Dorogovetsev S N, Goltsev A V and Mendes J F F, Rev. Mod. Phys. **80** (2008) 1275.
Liu W, Schmittmann B and Zia R K P 2012 *Europhys. Lett.***100** 66007.
Liu W, Jolad S, Schmittmann B and Zia R K P 2013 *J. Stat. Mech. Theory Exp.***2013** P08001.
Liu W, Schmittmann B and Zia R K P 2014 *J. Stat. Mech. Theory Exp.***2014** P05021.
Newman M E J, J. Stat. Phys., **101** (2000) 819.
Sen P, Dasgupta S, Chatterjee A, Sreeram P A, Mukherjee G and Manna S S, Phys. Rev. **E 67**, (2003) 036106.
Tyson J J, Chen K C and Novak B, *Current Opinion in Cell Biology*, (2003) , Elsevier.

Chapter 14
Social Physics: Understanding Human Sociality in Communication Networks

Asim Ghosh, Daniel Monsivais, Kunal Bhattacharya and Kimmo Kaski

Abstract In this brief review, we discuss some recent findings of human sociality in contemporary techno-social networks of interacting individuals. Here we will focus on a few important observations obtained by analysing mobile communication data of millions of users in a European country participating in billions of calls and text messages over a period of one year. In addition to the description of the basic structure of the network in terms of its topological characteristics like the degree distribution or the clustering coefficient, the demographic information of the users have been utilized to get deeper insight into the various facets of human sociality related to age and gender as reflected in the communication patterns of users. One of the observations suggests that the grandmothering effect is clearly visible in these communication patterns. In addition it is found that the number of friends or connections of a user show a clear decaying trend as a function of the user's age for both genders. Furthermore, an analysis of the most common location of the users shows the effect of distance on close relationships. As computational analysis and modelling are the two key approaches or tools of modern 'Social Physics' we will very briefly discuss the construction of a social network model to get insight into how plausible microscopic social interaction processes translate to meso- and macroscopic socially weighted network structures between individuals.

A. Ghosh (✉) · D. Monsivais · K. Bhattacharya · K. Kaski
Department of Computer Science, Aalto University School of Science,
P.O.Box 15500, 00076 Aalto, Finland
e-mail: asim.ghosh@aalto.fi

D. Monsivais
e-mail: daniel.monsivais@aalto.fi

K. Bhattacharya
e-mail: kunal.bhattacharya@aalto.fi

K. Kaski
e-mail: kimmo.kaski@aalto.fi

F. Abergel et al. (eds.), *Econophysics and Sociophysics: Recent Progress and Future Directions*, New Economic Windows,
DOI 10.1007/978-3-319-47705-3_14

14.1 Introduction

Recently, large amount of research on social communication has been done by using Big Data or records of "digital footprints" available from modes of modern-day communication such as mobile phone calls and text messages as well as social media like Twitter and Facebook. The reason for this is that it offers a complementary, easily measurable and quantifiable way to investigate social interactions between large number of people forming a network, in which people and social interactions correspond to its nodes and weighted links, respectively. The mobile phone communication data has turned out to be the first and most prominent so far in helping us to understand the microscopic details of social networks, human mobility and their behavioural patterns (Blondel et al. 2015) as well as how these microscopic properties convert to macroscopic features. Many interesting observations were found by analysing mobile communication, for example, we now have quite a bit of understanding of a number of structural properties of human social networks, such as the degree distribution, distribution of tie strengths, clustering coefficient, community structure, motif statistics, etc. (Onnela et al. 2007a, b; Kovanen et al. 2011; Tibély et al. 2011). A natural follow-up of data analysis or reality mining approach is modelling, to explore plausible mechanisms that reproduce observed structures and properties of the network (Kumpula et al. 2007; Toivonen et al. 2009). The combination of these two approaches, namely analysis and modelling, constitutes the modern empirical research approach of Social Physics, which is a concept coined by philosopher August Comte during the era of Industrial Revolution in the early 19th century while considering that the behaviour and functions of human societies could be explained in terms of underlying laws like in Physics.

Apart from the static structural properties influencing the functioning of social networks, the fact is that the dynamics therein and on them constitute another interesting area of network properties, i.e. social networks are temporal in nature. For example, there has been studies investigating inhomogeneous temporal sequences of communication events by looking at the distribution of the number of events in a bursty period that follows a universal power-law (Karsai et al. 2012). As a consequence, it was found that the spreading processes like rumour propagation become slow due to temporal inhomogeneity in event sequences or large inter-event times separating events, even though the networks have the small-world property (Karsai et al. 2011). It was also observed that such a heavy tail does not originate from circadian and weekly patterns of communicating users, rather it is a consequence of burstiness in temporal mobile phone communication patterns (Jo et al. 2012).

Apart from investigations of the basic properties of networks there has been quite a few studies using demographic data, by measuring gender differences in egocentric networks. The latter studies showed shifting patterns of communication from the reproductive age to the age of parental care (Palchykov et al. 2012; Karsai et al. 2012; Palchykov et al. 2013; David-Barrett et al. 2015). In addition, using most common location of the users in the data we have learned that the tie strength is related to the geographical distance (Onnela et al. 2011; Jo et al. 2014). Moreover, a universal pattern of time allocation to differently ranked alters has been found (Saramäki et al. 2014). By studying temporal motifs, homophily and gender specific communication patters were observed (Kovanen et al. 2013). A recent study also indicates variation of number of friends with the age and gender (Bhattacharya et al. 2015).

In this review, we will discuss a few important observations obtained by analysing a mobile communication data of millions of users participating in billions of calls and text messages in a year. Next we will provide detailed information of the data and methodologies used for the analysis. Following that we will mention fundamental observations from this data. First, we will discuss the relevance of the age distribution of the most frequently contacted person and its connection to the well known 'grandmothering hypothesis' (Hawkes et al. 1998). Second we will discuss a recent observation of the variation in the number of connections with the age and gender of the users. This is followed by a brief discussion of the effect of geographical distance on close relationships by considering the most common locations of the users. Then we will briefly discuss the construction of a social network model for exploring the mechanisms that produce the observed structures and properties of the network. Finally, we will end the review with a brief general discussion and some concluding remarks.

14.2 Data and Methods

The studied dataset contains anonymised mobile phone call detail records (CDRs) from a particular operator in a European country during 2007. The dataset contains full calling histories for the users of this service provider, whom are termed 'company users' and the users of other service providers are called 'non-company users'. There are more than six million company users and more than 25 million non-company users appearing in the full one year period records (Palchykov et al. 2012; Bhattacharya et al. 2015).

The dataset also contains demographic information of the company users about their gender, age, zip code and most common location. The zip code is extracted from the billing address of the user and the most common location (in terms of latitude and longitude) is the location of the cell tower most used by the user. However, the most common location does not necessarily correspond to the zip code. By using the latitude and longitude, the geographic distance between two locations can be calculated. In the dataset, the number of users with complete demographic information is around three million (Jo et al. 2014).

In the data set, it was found some company users have multiple subscriptions under the same contract numbers. For such users determination of their real age and gender is difficult. The gender and age of such users were not considered. The age of each company user was recorded when the contract was signed. Therefore, the age of each user was increased by the number of years between user's contract year and 2007 (Palchykov et al. 2012; Bhattacharya et al. 2015).

14.3 Observations

To study the dynamics of closest relationships between each ego (the individual in focus) and his/her alters (the contacts of the individual), the alters are ranked in terms of total number of calls (or total amount of calling time) with the ego. In order to interpret the findings, we have assumed that top ranked alters around the same age and opposite sex of those of the egos' are considered to be the egos' partners or spouses. We have also assumed that when the age of the top ranked alters is about one generation apart from the egos' age, they are considered to be egos' children or parents, irrespective of sex. To illustrate these relationships a small sample portion of mobile communication based social network is shown in Fig. 14.1. Here the blue circles represent male and red circles female users, with numbers within the circles denoting the age of the user and the size of the circle being linearly dependent on the user's age, while grey circles denote missing age and gender information, and the frequencies of contacts are denoted by the thickness of the links (and by the number on the links) (Palchykov et al. 2012).

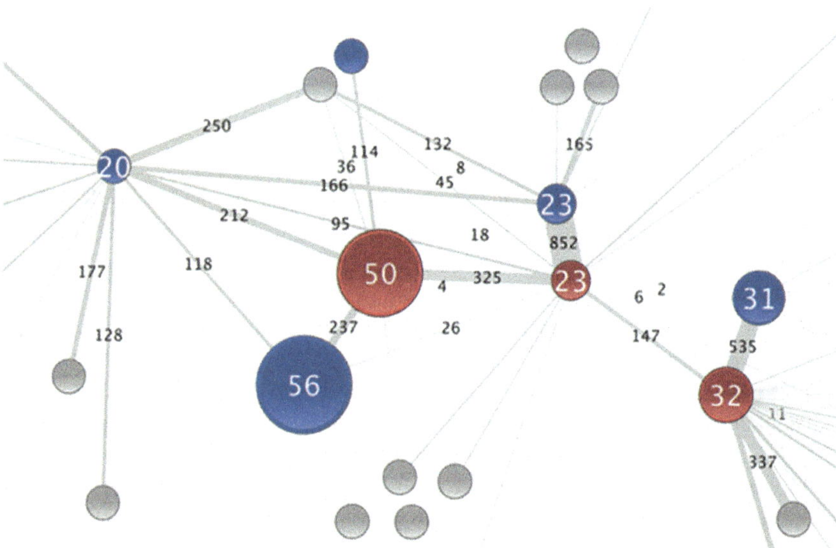

Fig. 14.1 A part of the network is shown. *Blue* and *red circles* represent male and female users, respectively. Also users' ages are denoted by circle sizes (older the age, bigger the circle) as well as numbers. Individuals for whom the age and the gender is not known are denoted by *grey circles*. Taken from (Palchykov et al. 2012)

14.3.1 Alter Age Distribution

The top ranked alter of a given ego is the one that the ego is most frequently in contact with, counting both the number of calls and text messages. Figure 14.2 shows the age distributions of the top ranked alters for both male and female egos aged 25 and 50 years. A bimodal distribution is observed for both genders peaking at around ego's own age and another peak appears at an age-difference of around 25 years, i.e. one generation apart. We have already mentioned that opposite-gender biased maxima at ego's own age correspond to the partners or spouses of the egos. On the other hand,

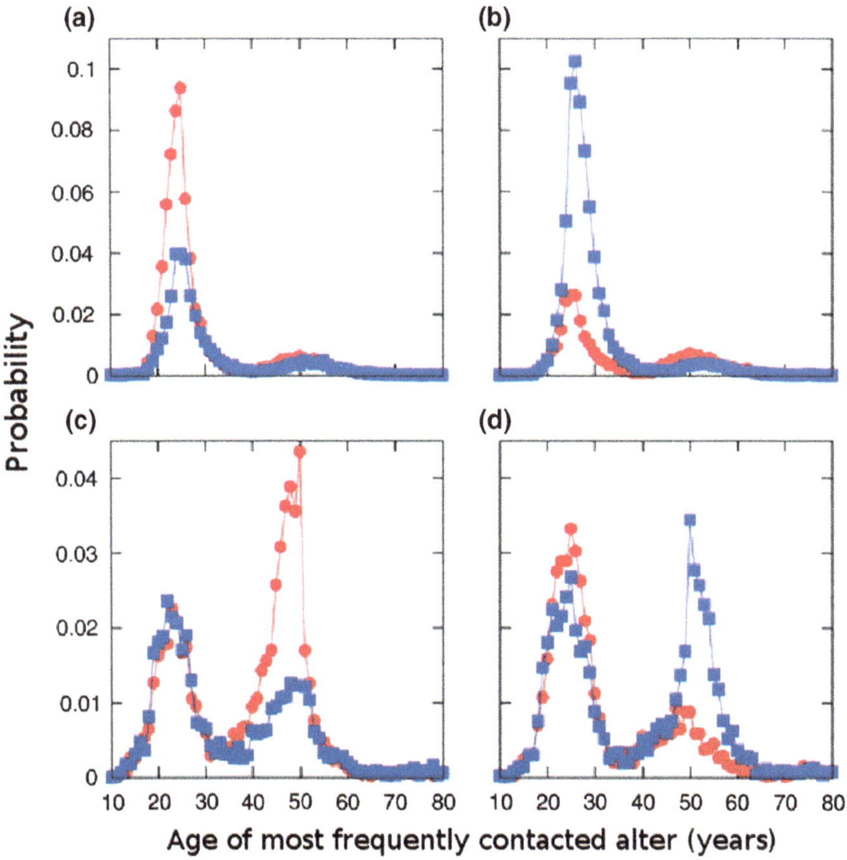

Fig. 14.2 The distributions of the highest ranked alters by age for 25 years old **a** male and **b** female egos. Figures (**c**) and (**d**) show similar distributions for 50 years old male and female egos, respectively. Red and blue circles represent female and male highest ranked alters, respectively. Each data point displays the probability that the highest ranked alters is of specified age and gender. Taken from (Palchykov et al. 2012)

the peaks at 25 year age difference from the ego's age could correspond children and parents, respectively, for 50 and 25 year old egos.

From Fig. 14.2, it has been observed that females are more focused on their partners during their reproductive period. Interestingly, when females are reaching the grandmothering age, they start to give more attention to their daughters than to their partners. The frequent connections between the mother and the daughter is a reflection of the grandmothering effect (Hawkes et al. 1998). Indeed, these observations point to the fact that females play an important role at their reproductive age as well as at the grandmothering age.

14.3.2 Variation in the Number of Alters with Egos' Age

In Fig. 14.3 we show the variation of the average number of egos' monthly contacts with alters as a function of the egos' age. In Fig. 14.3a we see that the number of alters reaches a maximum when egos are around 25 years old (Bhattacharya et al. 2015). Then the average number of alters decreases monotonically till the age of around 45 years for both male and female egos. From the age of 45 years onwards the average number of alters seems to stabilize for about 10 years for both male and female egos, but then again we see monotonous decrease after the egos' age of 55 years. The same behaviour is seen when we consider male and female egos separately, as depicted in

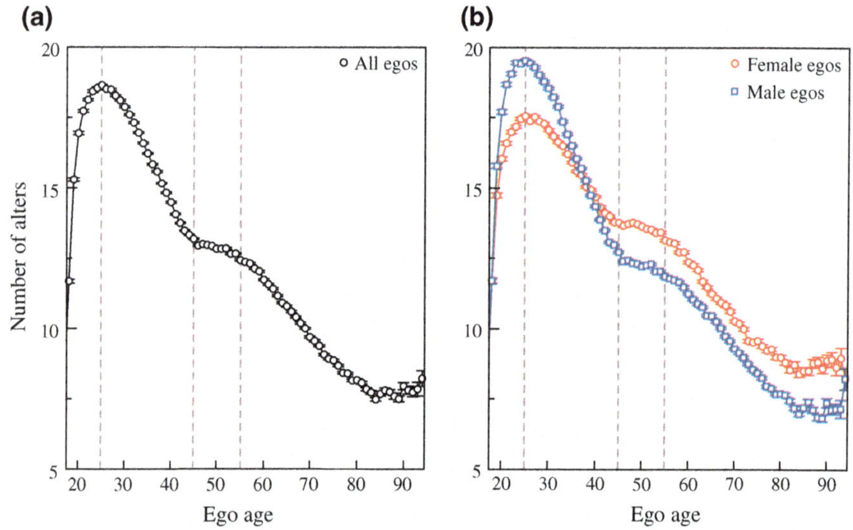

Fig. 14.3 The variation of the average number of egos' monthly contacts with alters as a function of the egos' age (years) for **a** all egos irrespective of their sexes, and **b** both sexes separately. *Blue* and *red squares* denote for male and female egos, respectively. The error bars span the 95 % confidence interval. Taken from (Bhattacharya et al. 2015)

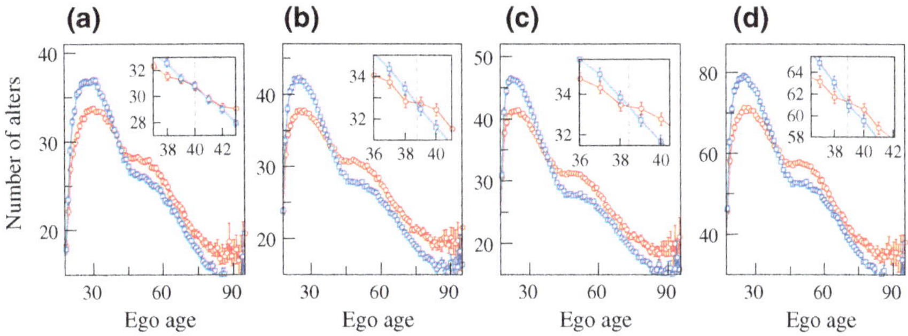

Fig. 14.4 The variations of the average number of alters as a function of egos age for different time windows of January-April, May-August and September-December are shown in (**a**), (**b**) and (**c**) respectively. The figure (**d**) is the result of time window of the whole year. The same figure legend is used as in Fig. 14.3b. The figures in the inset are used to focus crossover region. Taken from (Bhattacharya et al. 2015)

Fig. 14.3b. Also we would like to point out that the average number of alters for male egos is greater than that for female egos for egos' ages below 39 years and after that the number of alters for female egos turns out to be greater than that for male egos. The robustness of this finding is checked by taking different time windows as shown in Fig. 14.4. A consistent pattern is observed with the same crossover age at around 39 years, irrespective of the time window used.

14.3.3 Geographical Distance and Tie-Strength

In this section we discuss the correlations between the calling pattern and geographical separations by using the most common location of the egos and alters (Jo et al. 2014). The ego-alter pairs are split into four possible groups based on gender where M stands for male and F for female, namely F:F, M:M, F:M, and M:F. We have observed that the age distribution of the top-ranked alters is bimodal, with one peak around the ego's own age and another being one generation apart i.e., approximately 25 years apart. The current analysis is done by dividing all the ego-alter pairs into two categories where the age difference between egos and alters is $\leq$10 years and $>$10 years. In Fig. 14.5 the average fraction of alters living in a location different from that of the ego, termed geographic difference index (=1 if different and =0 if not), is shown as a function of the ego's age for the different gender and age groups.

The geographic difference indices for female and male egos turn out to be mostly identical for top-ranked alters of the opposite sex (F:M, M:F) as depicted in Fig. 14.5a. It is observed that the fraction of alters in a location different from that of the ego increases up to 0.7 for 20 years old egos, then it decreases to 0.45 by the mid-40's, and after that it remains approximately constant. A possible reason is that young couples live in the same neighbourhood before going to work or college, which would

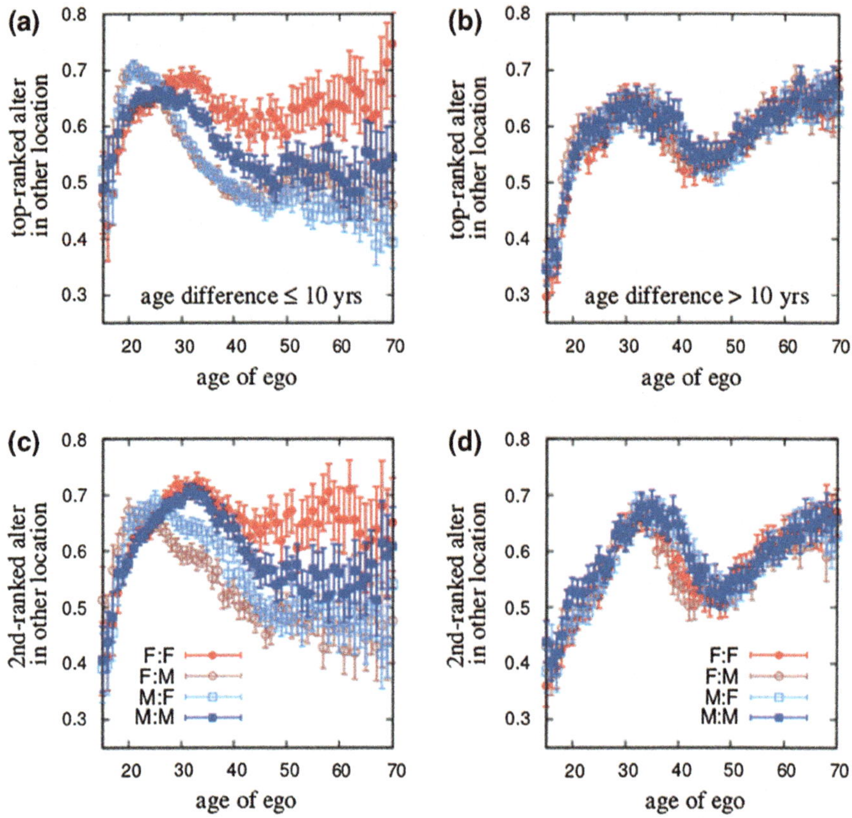

Fig. 14.5 The average fraction of alters living in a different location (municipality) of top-ranked alters (*top panels*) and 2nd-ranked alters to the ego (*bottom panels*). Ego-alter pairs with age difference less than 10 years is shown in *left panels*, while the age difference is larger than 10 years is shown in the *right panels*. "M:F" denotes male ego and female alter and so on. Error bars show the confidence interval with significance level $\alpha = 0.05$. Taken from (Jo et al. 2014)

suggest a larger distance. At older age, they eventually settle down together. However the geographic difference indices of ego-alter pairs with age difference ≤ 10 years behave differently and show gender dependence. The index slowly reaches the maximum around late 20s (M:M) or around 30 years (F:F), after which it decreases and fluctuates (M:M) or slightly decreases and increases again (F:F). After the maximum, the M:M curve remains lower than the F:F curve for all ages, an indication that the top-ranked male alters of male egos tend to live geographically closer.

For top-ranked alters with age difference >10 years, no significant change is observed based on the gender of egos and alters as shown in Fig. 14.5b. For all cases, at young age the egos and their top-ranked alters live in the same location, with probability $\approx 65\,\%$. The fraction of alters in a different location peaks in the 30s and after showing a local minimum at around mid-40s, it increases again. The minimum can be because of children living with their parents until leaving home in their young age. For older egos, their children would have already left their parental homes to live elsewhere contributing to an increase in the value of the index.

The index when measured with respect to the 2nd-ranked alters appears to be similar to that for the top-ranked alters, as is reflected in Fig. 14.5c–d. However, the peaks seem to appear at later ages. For age difference ≤ 10 years, the F:M and M:F curves are not overlapping and it appears that the index value is greater for the M:F curve compared to that of the F:M. This behaviour might appear from the partners of female egos being ranked 2nd more often than the partners or spouses of male egos. In particular females shift their interest from their partners to their children as they become older. For younger females, they are more likely to have a same-gender intimate friends than males are. For ego-alter pairs with age difference >10 years, the gender does not matter, except for females in their 30–40s whose 2nd-ranked alters are slightly more often located in the same place, compared to corresponding males. These 2nd rank alters of the older female egos could be their children (on the basis of the age difference between them) and the effect also lends support to the grandmothering hypothesis (Palchykov et al. 2012; Hawkes et al. 1998).

14.4 Related Modelling

We have so far been considering human sociality and structural features of social networks from the perspective of reality mining using data or making data-driven discoveries. However, in the toolbox of modern Social Physics we have also another key and complementary approach, namely computational modelling, which we will use here to get insight into how microscopic social interaction processes translate to meso- and macroscopic socially weighted network structures between individuals. As one of the first examples of this type of approach we refer to a rather simple model by Kumpula et al. (2007), which describes the processes for individuals getting acquainted with each others leading in turn to the formation of locally and globally complex weighted social network structures.

In this model one consider a network with a fixed number of N nodes, where links can be created in two ways: First, in a time interval Δt each node having at least one neighbour starts a weighted local search for new friends, see Fig. 14.6a, b. Then the node i chooses one of its neighbouring node j with probability w_{ij}/s_i, where w_{ij} represents the weight of the link connecting i with j and $s_i = \sum_j w_{ij}$ is the strength of node i. If the chosen node j has other neighbours k (apart from i), it chooses one of them with probability $w_{jk}/(s_j - w_{ij})$ implying that the search favours strong links. If there is no connection between i and k, it will be established with probability $p_\Delta \Delta t$ such that $w_{ik} = w_0$. If the link exists, then its weight will be increased by a certain amount δ. In addition, both w_{ij} and w_{jk} are increased by δ. This kind of cyclic closure mechanism of "friend of a friend will also be friend" corresponds to *local attachment* (LA). On the other hand if a node has no links, it will create a link of weight w_0 with probability $p_r \Delta t$ to a random node, as depicted in Fig. 14.6c. This mechanism is to

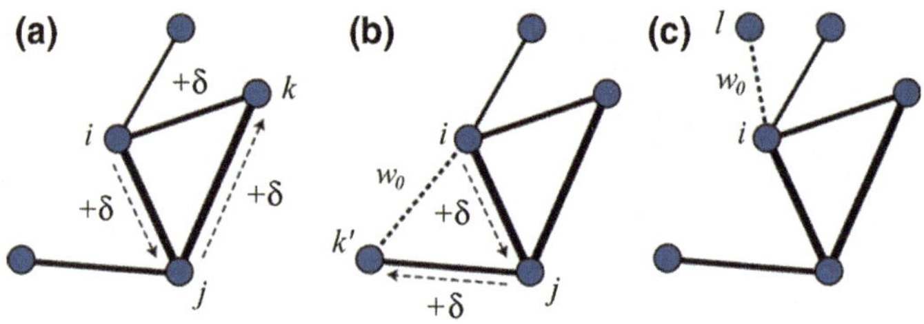

Fig. 14.6 Schematic diagram of the model algorithm. **a**: a weighted local search starts from i and proceeds to j and then to k, which is a neighbour of i also. **b**: the local search from i ends to k', which is not a neighbour of i. In this case link $w_{ik'}$ is set with probability p_Δ. **c**: node i creates a link randomly to a random node l with probability p_r. In cases (**a**) and (**b**) the involved links weights are increased by δ. Taken from (Kumpula et al. 2007)

establish a new link outside the immediate neighbourhood of the chosen node, like in focal closure mechanism, corresponding to *global attachment* (GA). In addition to these two basic link formation mechanisms the model introduces with probability $p_d \Delta t$ a *node deletion* (ND), in which all the links of a node are removed while the node itself is kept to maintain fixed system size.

The simulation runs are started with N nodes without any links, such that LA and GA mechanisms are updated in parallel followed by the ND step (assuming $\Delta t = 1$, $w_0 = 1$, $p_d = 10^{-3}$ and $p_r = 5 \times 10^{-4}$). The parameter δ is responsible for the time-dependent development of the weights of the network. In this study to observe the behaviour of the network for different δ values the simulation runs were performed such that the average degree was kept fixed ($\langle k \rangle \approx 10$). For each δ, the parameter p_Δ was adjusted to keep $\langle k \rangle$ constant. These simulations were performed for four values of $\delta = 0, 0.1, 0.5, 1$ as shown in Fig. 14.7. Here $\delta = 0$ implies the unweighted networks. On the other hand for higher values of δ the obtained network structures show clearly the formation of communities, due to favouring the LA mechanism. This helps to follow same links simultaneously which in turn leads to the increased link weights and associated triangles. So in the steady state, any triangle starts to rapidly accumulate weight and contribute to the formation of weighted network.

The results in Fig. 14.7, with increasing δ values show the emergence of communities and specifically for larger δ values one sees a community structure very similar to that observed in reality mining studies of a mobile phone dataset by Onnela et al. (2007a, b). In further analysis it becomes evident that this rather simple model is able to reproduce a number of stylised facts of social network found in these empirical studies. Most importantly this model is able to show the same local and global structure giving further verification of the Grannovetter's "strength of weak ties" hypothesis stating that "The stronger the tie between A and B, the larger the proportion of individuals S to whom both are tied" (Granovetter 1973). This in turn indicated that the triadic and focal closure mechanisms, as proposed by Kossinets and Watts (2006), are at least plausible if not the most important mechanisms playing role in the formation of a social network.

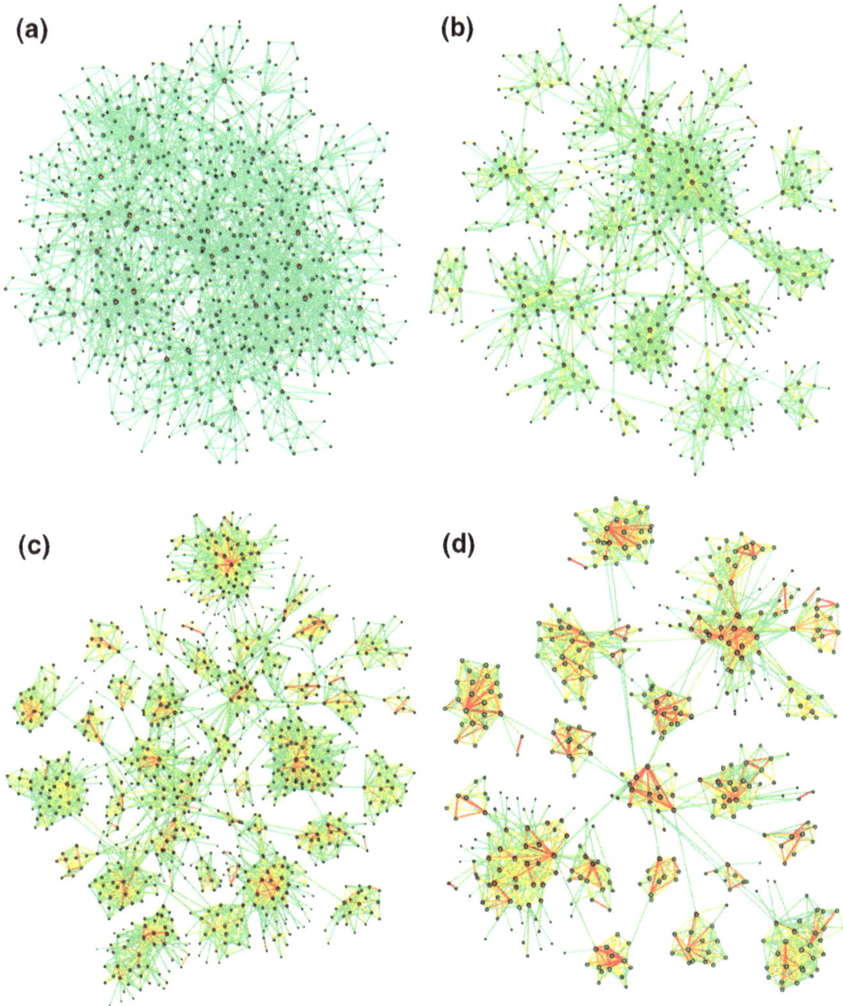

Fig. 14.7 Networks with **a** $\delta = 0$, **b** $\delta = 0.1$, **c** $\delta = 0.5$, and **d** $\delta = 1$. Link colours change from green (weak links) to *yellow* and *red* (strong links). Taken from (Kumpula et al. 2007)

14.5 Discussion and Conclusions

Apart from the face to face communication, other communication modalities especially those based on mobile devices have become increasingly important in our social lives, yet serving as means of expressing emotional closeness between two individuals, which is generally reflected as the strength or frequency of communication between them. This is well exemplified in Fig. 14.1 where a small portion of mobile communication based proxy of a social network is shown. In studies of mobile communication patterns, we assume based on demographic information of the service subscribers that top ranked alters close to the same age group and opposite genders can be considered as the egos' partners or spouses. We further assume that when the age of the top ranked alters is one generation apart from the egos' age, they are egos' children or parents, irrespective of their gender.

By measuring the age distribution of top ranked, i.e. most frequently contacted alters for given age group of the egos, we have observed that females are more focused on their partners during their reproductive period (Fig. 14.2). Interestingly we also observe that when females are reaching the grandmothering age, they start to give more attention to their children with emphasis on daughters than to their partners. Such frequent contacts between the mother and daughter reflect the grandmothering effect. Hence all these observations tells us that females play an important role in our society not only at their reproduction age but also at their grandmothering age.

From Fig. 14.3, we have observed that the maximum number of connections for both males and females occur at the age of around 25 (Bhattacharya et al. 2015). During this younger age, males are first found to be more connected than females. After the age of 25 years, the number of alters decreases steadily for both males and females, although the decay is faster for the males than for the females. These different decay rates result in a crossover around the age of 39 years when females become more connected than males. Note, that for the age from 45 to 55, the number of alters stabilizes for both males and females. This age cohort is the one in which the egos' children typically marry and begin to reproduce. Therefore, one likely explanation for this plateau from age 45 to 55 is that it reflects the case that parents are maintaining regular interaction with their children at a time when some of these might otherwise be lost. The gap between the sexes seems to be primarily due to the more frequent interactions by the mothers with their adult children and the children's spouses. Also it was found that females interact with their own close family members and the new in-laws formed by their children's marriages more than males do.

By using the most common geographical location of the users available in mobile communication dataset, one gets insight into their age and gender dependent life-course migration patterns (Jo et al. 2014). The analysis suggests that young couples tend to live further apart from each other than old couples (Fig. 14.5). Also it was found using the post code information in the dataset that emotionally closer pairs are living geographically closer to each other.

In the end of this brief report we have discussed a simple model for a social network formation, where the link weights between a pair of individuals and their dynamics were considered to be the essential ingredients to give us insight how these microscopics translate to meso- and macroscopic structural properties of the societal scale network (Kumpula et al. 2007). In the model, the coupling between network structure and social interaction strengths is established by the link weights between individuals. A new link is added or strengthened depending on the existing weights. In this process, the communities will emerge when nodes are sufficiently strong to connecting new ones. This rather simple model turned out to reproduce many of the stylised facts found in empirical studies of weighted social networks, e.g. by verifying Granovetter's "strength of weak ties" hypothesis and giving further evidence that triadic and focal closure are the two key mechanisms in explaining the formation of communities in social networks.

Acknowledgements A.G. and K.K. acknowledge support from project COSDYN, Academy of Finland (Project no. 276439). K.B., D.M. and K.K. acknowledge support from H2020 EU project IBSEN. D.M. acknowledge to CONACYT, Mexico for supporting grant 383907.

References

K. Bhattacharya, A. Ghosh, D. Monsivais, R. I. Dunbar, and K. Kaski, "Sex differences in social focus across the lifecycle in humans," *Royal Society Open Science*, vol. 3, p. 160097, 2015.

V. D. Blondel, A. Decuyper, and G. Krings, "A survey of results on mobile phone datasets analysis," *EPJ Data Science*, vol. 4, no. 1, pp. 1–55, 2015.

T. David-Barrett, J. Kertesz, A. Rotkirch, A. Ghosh, K. Bhattacharya, D. Monsivais, and K. Kaski, "Communication with family and friends across the life course," *arXiv preprints* arXiv:1512.09114, 2015.

M. S. Granovetter, "The strength of weak ties," *American journal of sociology*, pp. 1360–1380, 1973.

K. Hawkes, J. F. OConnell, N. B. Jones, H. Alvarez, and E. L. Charnov, "Grandmothering, menopause, and the evolution of human life histories," *Proceedings of the National Academy of Sciences*, vol. 95, no. 3, pp. 1336–1339, 1998.

H.-H. Jo, M. Karsai, J. Kertész, and K. Kaski, "Circadian pattern and burstiness in mobile phone communication," *New Journal of Physics*, vol. 14, no. 1, p. 013055, 2012.

H.-H. Jo, R. I. Dunbar, J. Saramäki, and K. Kaski, "Dynamics of close relationships for the life-course migration," *Scientific Reports*, vol. 4, p. 6988, 2014.

M. Karsai, K. Kaski, A.-L. Barabási, and J. Kertész, "Universal features of correlated bursty behaviour," *Scientific reports*, vol. 2, 2012.

M. Karsai, M. Kivelä, R. K. Pan, K. Kaski, J. Kertész, A.-L. Barabási, and J. Saramäki, "Small but slow world: How network topology and burstiness slow down spreading," *Physical Review E*, vol. 83, no. 2, p. 025102, 2011.

M. Karsai, K. Kaski, and J. Kertész, "Correlated dynamics in egocentric communication networks," *Plos one*, vol. 7, no. 7, p. e40612, 2012.

G. Kossinets and D. J. Watts, "Empirical analysis of an evolving social network," *science*, vol. 311, no. 5757, pp. 88–90, 2006.

L. Kovanen, M. Karsai, K. Kaski, J. Kertész, and J. Saramäki, "Temporal motifs in time-dependent networks," *Journal of Statistical Mechanics: Theory and Experiment*, vol. 2011, no. 11, p. P11005, 2011.

L. Kovanen, K. Kaski, J. Kertész, and J. Saramäki, "Temporal motifs reveal homophily, gender-specific patterns, and group talk in call sequences," *Proceedings of the National Academy of Sciences*, vol. 110, no. 45, pp. 18070–18075, 2013.

J. M. Kumpula, J.-P. Onnela, J. Saramäki, K. Kaski, and J. Kertész, "Emergence of communities in weighted networks," *Physical review letters*, vol. 99, no. 22, p. 228701, 2007.

J.-P. Onnela, J. Saramäki, J. Hyvönen, G. Szabó, D. Lazer, K. Kaski, J. Kertész, and A.-L. Barabási, "Structure and tie strengths in mobile communication networks," *Proceedings of the National Academy of Sciences*, vol. 104, no. 18, pp. 7332–7336, 2007a.

J.-P. Onnela, J. Saramäki, J. Hyvönen, G. Szabó, M. A. De Menezes, K. Kaski, A.-L. Barabási, and J. Kertész, "Analysis of a large-scale weighted network of one-to-one human communication," *New Journal of Physics*, vol. 9, no. 6, p. 179, 2007b.

J.-P. Onnela, S. Arbesman, M. C. González, A.-L. Barabási, and N. A. Christakis, "Geographic constraints on social network groups," *PLoS one*, vol. 6, no. 4, p. e16939, 2011.

V. Palchykov, K. Kaski, J. Kertész, A.-L. Barabási, and R. I. Dunbar, "Sex differences in intimate relationships," *Scientific reports*, vol. 2, 2012.

V. Palchykov, J. Kertész, R. Dunbar, and K. Kaski, "Close relationships: A study of mobile communication records," *Journal of Statistical Physics*, vol. 151, no. 3-4, pp. 735–744, 2013.

J. Saramäki, E. A. Leicht, E. López, S. G. Roberts, F. Reed-Tsochas, and R. I. Dunbar, "Persistence of social signatures in human communication," *Proceedings of the National Academy of Sciences*, vol. 111, no. 3, pp. 942–947, 2014.

G. Tibély, L. Kovanen, M. Karsai, K. Kaski, J. Kertész, and J. Saramäki, "Communities and beyond: mesoscopic analysis of a large social network with complementary methods," *Physical Review E*, vol. 83, no. 5, p. 056125, 2011.

R. Toivonen, L. Kovanen, M. Kivelä, J.-P. Onnela, J. Saramäki, and K. Kaski, "A comparative study of social network models: Network evolution models and nodal attribute models," *Social Networks*, vol. 31, no. 4, pp. 240–254, 2009.

Chapter 15
Methods for Reconstructing Interbank Networks from Limited Information: A Comparison

Piero Mazzarisi and Fabrizio Lillo

Abstract In this chapter, we review and compare some methods for the reconstruction of an interbank network from limited information. By exploiting the theory of complex networks and some ideas from statistical physics, we mainly focus on three different methods based on the maximum entropy principle, the relative entropy minimization, and the fitness model. We apply our analysis to the credit network of electronic Market for Interbank Deposit (e-MID) in 2011. In comparing the goodness of fit of the proposed methods, we look at the topological network properties and how reliably each method reproduces the real-world network.

15.1 Introduction

Starting from the subprime mortgages crisis of 2007–2009, a growing interest is focused on the problem of assessing the systemic risk of a financial system. Among the many different aspects of systemic risk studied by the recent literature, the study of financial networks is widely recognized as one of the most important. Indeed, between financial agents operating in a financial system, there are typically a large number of different reciprocal ties, e.g. credit relations, equity investments, securities, exposures, to name but a few, which create a highly connected structure with the features of a complex network (Bonanno et al. 2003; Billio et al. 2012; Gai and Kapadia 2010) or a multiplex (Bargigli et al. 2015). In this contest, systemic risk refers to the propagation of financial distresses as a consequence of the connections between financial agents. Regarding the specific case of interbank markets, banks are represented as the nodes of the financial network and credit relations are represented by links. Liquidity shocks may propagate among the interconnected banks.

P. Mazzarisi (✉) · F. Lillo
Scuola Normale Superiore, Piazza dei Cavalieri 7, 56126 Pisa, Italy
e-mail: piero.mazzarisi@sns.it

F. Lillo
e-mail: fabrizio.lillo@sns.it

F. Abergel et al. (eds.), *Econophysics and Sociophysics: Recent Progress and Future Directions*, New Economic Windows,
DOI 10.1007/978-3-319-47705-3_15

Several new and old measures of systemic risk try to capture this phenomenon (Cont et al. 2010; Battiston et al. 2012; Caccioli et al. 2014). However, the analysis of systemic risk for the banking system might be restricted due to the unavailability of full information on the network structure which causes the data to be limited. Network theory and statistical physics may provide a solution by making possible to reconstruct details of the financial network from partial sets of information.

In this paper we review and apply some of the recently proposed methodologies to the case of the electronic market for Italian overnight interbank lending (e-MID). Before starting our review, we note that the objective of network reconstruction can be twofold. The first approach to network reconstruction aims at reproducing the topological features of real-world credit network with limited information about banks' balance sheet. This means predicting the presence of an interbank relation, the number of counterparts of a bank, etc. In network jargon, this approach studies the problem of inferring the probability of a link from limited information on the network. The second approach does not necessarily consider as an objective the faithful reconstruction of the network, but rather of the systemic risk of each bank (Mastromatteo et al. 2012; Di Gangi et al. 2015). In this sense a good reconstruction could also give relatively biased or incorrect topological reconstruction if the systemic risk metrics of the nodes of the reconstructed network are close to those of the real network. In this review, we focus on the first type of approach.

15.2 Interbank Lending and the E-MID Market

The interbank lending market is a market in which banks extend loans to one another for a specified term and/or collateral. A significant fraction of interbank loans are for maturities of one week or less, the majority being overnight.

In inferring systemic risk of a credit network, there is a major problem: lack of complete information about network structure of the market. Depending on the requirements imposed by central banks and other regulatory bodies, all banks in the interbank market must declare the total lending volume and the total borrowing volume, but the information about the presence of a deposit between two banks is often unavailable. In mathematical words, given the weighted adjacency matrix[1] representing the credit network, see the left top panel of Fig. 15.1, one has access to the marginals $\{a_i\}$, $\{l_j\}$, representing the total interbank exposures in the asset side and in the liability side, but there is no information about the value of the generic weight ω_{ij}, that represents the amount of loan from bank i to bank j. This is the starting point of every method of credit network reconstruction and from now on we assume to know at least the marginals of the weight matrix representing the interbank lending market. Other information can be available and used for the reconstruction.

[1]The weighted adjacency matrix or weights matrix is the generalization of the adjacency matrix in the case of weighted (directed) graphs.

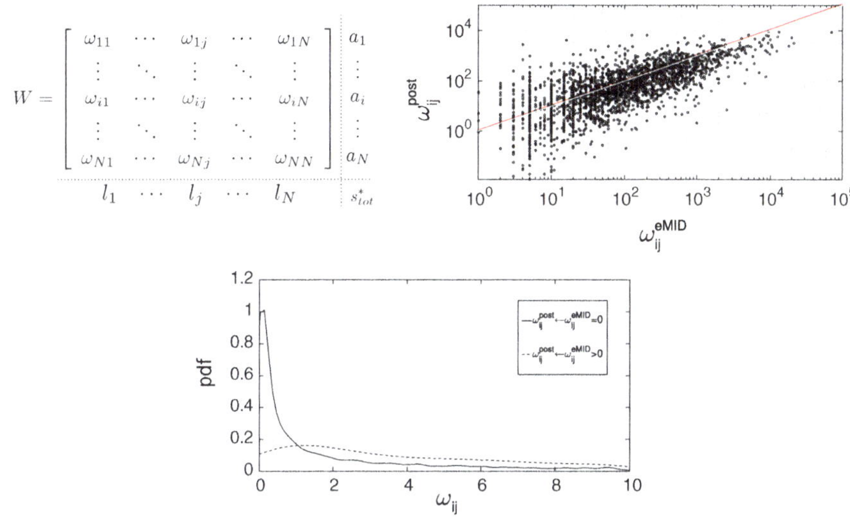

Fig. 15.1 *Left top panel* weight matrix of interbank lending exposure for a credit market. *Right top panel* scatter plot of the weights reconstructed with the relative entropy and the real non vanishing ones. *Bottom panel* density estimation of the weights reconstructed with the relative entropy method, considering separately the real weights equal to zero and different from zero

As testing dataset we adopt the e-MID credit network. The e-MID is an electronic market for Interbank Deposits in the Euro Area and it was founded in Italy in 1990 for Italian Lira transactions and denominated in Euros in 1999. The ever-increasing number of international counterparts joining e-MID confirms its leading role in liquidity pooling. According to the "Euro Money Market Study 2006 published by the European Central Bank in February 2007, e-MID accounted for 17 % of total turnover in unsecured money market in the Euro Area. More recently the amount of overnight lending in e-MID has significantly declined, especially around the sovereign debt crisis (Barucca and Lillo 2015). In this paper, we focus on the overnight deposits including both Italian and foreign banks. For further information about the network features of the Italian overnight money market see (Iori et al. 2008). The dataset contains the edge list of all credit transactions in each day of the year 2011. Throughout the paper, weights are expressed in millions of Euro.

15.3 Relative Entropy Minimization with Prior

In this Section, we analyze a classical method of network reconstruction introduced for the first time in 1991 by (Blien and Graef 1991) and widely adopted in estimating systemic risk for interbank markets before the subprime mortgage crisis, see for example (Sheldon et al. 1998; Upper and Worms 2004; Wells 2004). Also after the crisis, it represents the common approach in financial industry and is usually adopted in financial literature as null model in order to have a well-known benchmark, see (Mistrulli 2011; Silvestri and Cont 2015).

The method we present in this Section is based on the assumption that banks maximize the dispersion of their interbank exposures. Mathematically it consists in the minimization of relative entropy[2] with respect to a prior. This approach is theoretically supported by the work of Allen and Gale (Allen and Gale 2000) according to which a complete network structure makes the credit market more robust than an incomplete structure in the case of distress propagation. In fact, the typical output of relative entropy minimization is a complete network structure, that is a fully-connected network where each bank-node is connected to all the others.

However, real-world credit network are sparse and sparsity is recognized as one of the crucial features which characterize the phenomenon of financial contagion (Gai and Kapadia 2010). In the next Section, we will compare the present method with others which are not affected by this problem.

The method of reconstructing a credit network via relative entropy minimization is faced by solving the problem of adjusting the entries of a large matrix to satisfy prior consistency requirements. This problem is called Matrix Balancing Problem and can be solved by several matrix scaling algorithms. A widely used algorithm is the one called RAS. It was proposed for the first time by the Leningrad architect G.V. Sheleikhovskii for calculating passenger flow. Bregman (1967) proved that if an allowable solution of the (following) problem exists, then the RAS algorithm converges to the *optimal* solution.

The problem can be stated in the following way: the interbank linkages are represented by a $n \times n$ matrix $W = \{\omega_{ij}\}$, where $a_i = \sum_{j \neq i} \omega_{ij} \equiv s_i^{out}$ and $l_j = \sum_{i \neq j} \omega_{ij} \equiv s_j^{in}$ are, respectively, the total amount of money bank i lends to other banks and bank j borrows from other banks. In network jargon they are known as out- and in-strength respectively. By removing the possibility of self-loops (i.e. setting $\omega_{ii} = 0$, $\forall i$, we have to estimate $n^2 - n$ unknowns. The main goal is to estimate the entries of the matrix W which minimizes the relative entropy

$$\min_{\omega_{ij}} \sum_{i,j \neq i} \omega_{ij} \log \frac{\omega_{ij}}{p_{ij}}$$
$$\text{s.t.} \sum_{j \neq i} \omega_{ij} = a_i \ , \sum_{i \neq j} \omega_{ij} = l_j \ , \quad \omega_{ij} \geq 0 \ \ \forall i, j \ , \quad \omega_{ii} = 0 \tag{15.1}$$

where p_{ij} is a matrix representing a prior bias based on the assumption that banks maximize the dispersion of their interbank exposures. This implies that the prior is

$$p_{ij} = \frac{a_i l_j}{s_{tot}} \quad \forall i \neq j \tag{15.2}$$

[2]Relative entropy is also known as cross-entropy or Kullback-Leibler divergence.

where $s^*_{tot} = \sum_i a_i = \sum_j l_j$ is the total strength. To avoid self-loops, we set $p_{ij} = 0$ $\forall i = j$. By construction, the prior matrix does not fulfil the constraints on the strength sequences but allows *a priori* to set some entries of the network W equal to zero.

The RAS algorithm (Bacharach 1965) is a iterative proportional fitting procedure based on the rescaling of the prior's entries and is formally defined as follows:

Inizialization At $t = 0 \longrightarrow \omega^{(t)}_{ij} = p_{ij}\ \forall i, j$.
row scaling Let us define $\rho^{(t)}_i = \frac{a_i}{\sum_j \omega^{(t)}_{ij}}\ \forall i$
and update $\omega^{(t)}_{ij} \leftarrow \rho^{(t)}_i \omega^{(t)}_{ij}\ \forall i, j$.
column scaling Let us define $\sigma^{(t)}_j = \frac{l_j}{\sum_i \omega^{(t)}_{ij}}\ \forall j$
and update $\omega^{(t)}_{ij} \leftarrow \sigma^{(t)}_j \omega^{(t)}_{ij}\ \forall j, i$.
Iteration Let us set $t \leftarrow t + 1$ until the desired precision.

This approach is computationally efficient and the solution describes a fully-connected network according to the idea of fully-diversified interbank exposures. Since a credit market like e-MID is represented by a sparse network, we can expect that this approach does not reproduce well the characteristics of the real world.

In the right top panel and in the bottom panel of Fig. 15.1, we compare the weights reconstructed by the relative entropy with those in the real network. The right top panel focuses on the weights different from zero in the real network and shows that (i) a significant dispersion is observed for intermediate values of the real weights and (ii) the reconstructed weights are generically underestimated for large real weights. The bottom panel compares the probability distribution of reconstructed weights considering separately the case when the real weights are zero (i.e. no link) and different from zero. Correctly the former distribution is peaked close to zero, however the latter is very broad and it might be problematic to choose which reconstructed values should be set to zero.

Finally in order to compare this approach with the ones presented below, let us notice that the posterior solution can be also obtained by the method of Lagrange multipliers applied to Eq. (15.1):

$$\frac{\partial}{\partial \omega_{ij}} \left\{ \sum_{i,j\neq i} \omega_{ij} \log \frac{\omega_{ij}}{p_{ij}} - \sum_i \mu^{out}_i \left(a_i - \sum_{j\neq i} \omega_{ij} \right) - \sum_j \mu^{in}_j \left(l_j - \sum_{i\neq j} \omega_{ij} \right) \right\} = 0. \tag{15.3}$$

By solving ad defining

$$\phi^{out}_i \equiv \frac{a_i}{\sqrt{s^*_{tot}}} e^{-(\mu^{out}_i + \frac{1}{2})}, \quad \psi^{in}_j \equiv \frac{l_j}{\sqrt{s^*_{tot}}} e^{-(\mu^{in}_j + \frac{1}{2})}, \tag{15.4}$$

the entries of the posterior are simply equal to

$$\omega^*_{ij} = \phi^{out}_i \psi^{in}_j. \tag{15.5}$$

Below we will compare the solution of Eq. (15.5) with exponential random graphs.

15.4 Reconstructing via Exponential Random Graphs

In this Section, we review the problem of network reconstruction by exploiting statistical network models. In the previous case, the output was the posterior which represents the closer network to the real one according to the relative entropy. Here the aim is to obtain a probability distribution over the weight matrix which describes an ensemble of networks having specific characteristics we want to take fixed on average, (i.e. *canonical* ensemble). In other words, we know some marginals, e.g. total interbank exposures, and we ask for the probability of a link and the weight associated with it.

This objective is achieved by choosing the probability distribution for the network ensemble according to the principle of maximum entropy (ME) (Park and Newman 2004), that is the probability distribution maximizing the Shannon-Gibbs entropy. Once the specific marginals are chosen and the probability distribution formally specified, the maximum-likelihood (ML) method (Garlaschelli and Loffredo 2008) can be successfully applied for estimating the network model. The class of network models obtained through this approach is called exponential random graphs.

In this Section, we study the so-called *bosonic* graphs[3] obtained by fixing the strength sequences, i.e. banks' assets and liabilities. Similarly to the previous method, the bosonic model is not able to capture the sparsity of the real networks. Therefore, we present a generalization (Bargigli 2014) which in addition considers the total number of links as known. In principle, this quantity might not be available. However, looking at the data, the total number of links is quite constant in time, especially at the daily and weekly aggregated timescale. In other words we do not address the problem of estimating the number of links (L) of the real-world network but in our analysis we assume that this information is known. We want to test how much this additional information improves the network reconstruction,

Let us define the indicator function $\Theta \equiv \Theta(\omega)$ as the function taking value equal to one if $\omega > 0$, otherwise being zero. The degree of a node is the number of counterparts for a bank in the considered time window, specifically the out-degree $k_i^{out} = \sum_{j \neq i} \Theta(\omega_{ij})$ is the number of counterparties bank i is lending to, while the in-degree $k_j^{in} = \sum_{i \neq j} \Theta(\omega_{ij})$ is the number of counterparties bank j is borrowing from. Finally, the total number of links is simply $L = \sum_i \sum_{j \neq i} \Theta(\omega_{ij})$.

15.4.1 *Maximum Entropy Probability Distribution*

In this Subsection, we briefly review the general framework of exponential random graphs. Further information can be found in (Park and Newman 2004). The main goal is obtaining the generic probability distribution for the ensemble of graphs reproducing on average the marginals. Let $G \in \mathcal{G}$ be a graph in the ensemble with

[3] The word bosonic is used for the analogy with the Bose gas in Physics since the probability distribution for the considered model turns out to be the Bose-Einstein statistics.

N nodes and let $P(G)$ be the probability associated with the same graph within the ensemble. We choose $P(G)$ such that the expectation values of the observables, $\{< C_i(G) >_{G\in\mathcal{G}}\}$[4] are equal to their observed values $\{C_i^*\}$.

$P(G)$ is chosen by maximizing the Shannon-Gibbs entropy

$$S = -\sum_{G\in\mathcal{G}} P(G)\ln P(G) \;\; s.t. \;\; \sum_{G\in\mathcal{G}} P(G)C_i(G) = C_i^*, \;\; \sum_{G\in\mathcal{G}} P(G) = 1. \tag{15.6}$$

The quantity S defined in Eq. (15.6) is the one that best represents the lack of information beyond the known observables. By solving this problem, we obtain a network probability distribution in which no redundant information is considered.

By introducing the Lagrange multipliers α, $\{\theta_i\}$, the maximum entropy probability distribution is the one solving the following functional equation

$$\frac{\partial}{\partial P(G)}\left\{S + \alpha\left(1 - \sum_{G\in\mathcal{G}} P(G)\right) + \sum_i \theta_i\left(C_i^* - \sum_{G\in\mathcal{G}} P(G)C_i(G)\right)\right\} = 0. \tag{15.7}$$

Formally, the solution is

$$P(G,\boldsymbol{\theta}) = \frac{e^{-H(G,\boldsymbol{\theta})}}{Z(\boldsymbol{\theta})}, \quad \text{with} \quad H(G,\boldsymbol{\theta}) = \sum_i \theta_i C_i(G), \quad Z(\boldsymbol{\theta}) = e^{\alpha+1} = \sum_{G\in\mathcal{G}} e^{-H(G,\boldsymbol{\theta})}. \tag{15.8}$$

where $H(G)$ is the graph Hamiltonian and $Z(\boldsymbol{\theta})$ is the partition function. Eq. (15.8) define the exponential random graphs ensemble.

The estimation of the network model requires to find the values of the parameters θ_i which satisfy

$$\langle C_i\rangle_{\boldsymbol{\theta}^*} = C_i(G^*) \;\; \forall i \tag{15.9}$$

Garlaschelli and Loffredo (Garlaschelli and Loffredo 2008) proved that this method is statistically rigorous, in the sense that leads to unbiased information, for exponential random graphs. Furthermore, they proved that the solution of the Eq. (15.9) is equivalent to the solution of the maximum likelihood problem, which consists in maximizing the likelihood associated with the real-world network G^*, that is $P(G^*|\boldsymbol{\theta})$. By maximizing the likelihood, the optimal choice for $\boldsymbol{\theta} \equiv \boldsymbol{\theta}^*$ yields a (unique[5]) parameters estimation. A deeper explanation of Maximum Likelihood Estimation (MLE) for exponential random graphs and extensive applications can be found in (Squartini and Garlaschelli 2011; Squartini et al. 2015). Another widespread method for estimating exponential family models was introduced in (Geyer and Thompson 1992) by adopting Markov Chain Monte Carlo (MCMC) methods and successively applied

[4] $C_i(G)$ is the value of the observable C_i associated with graph G of the ensemble. In the *microcanonical* ensemble, we choose $C_i(G)$ equal to the observed quantity C_i^* for each graph G in the ensemble. In the *canonical* ensemble, this equality holds only in average.

[5] In the grand canonical ensemble, the solution is unique except for a specific shift of the Lagrange multipliers (symmetry of the Hamiltonian of the network models).

to the specific case of exponential random graphs (Snijders 2002). For a comparison of the two estimation methods, MLE and MCMC, see (Van Duijn et al. 2009). In our analysis, we adopt MLE.

15.4.2 Bosonic Configuration Model and Sparse Generalization

Bosonic exponential random graphs are obtained by imposing as constraints the in- and out- strength sequence of nodes. Thus the ensemble is specified by the graph Hamiltonian

$$H_{bosonic}(W,\boldsymbol{\theta}) = \sum_i \{\theta_i^{out} s_i^{out} + \theta_i^{in} s_i^{in}\} = \sum_i \sum_{j \neq i} (\theta_i^{out} + \theta_j^{in})\omega_{ij}. \tag{15.10}$$

Since the Hamiltonian is linearly proportional to the sum of the weights ω_{ij}, the partition function in Eq. (15.8) can be analytically computed for the model and as a consequence the probability distribution in Eq. (15.8) of the network ensemble is obtained. Specifically, the probability for a graph is simply the product of $N(N-1)$ independent geometric distributions with parameters related to the Lagrange multipliers $\boldsymbol{\theta}$. See (Garlaschelli and Loffredo 2009) for further specifications of the bosonic network model, also known as bosonic configuration model.[6] In the following, we indicate this model also as Directed Weighted Network model with fixed Strength sequence ($DWNS$).

The sparse generalization of the DWNS is obtained by imposing also the total number of links as constraint. The network Hamiltonian is

$$\begin{aligned} H_{sparse\ bosonic}(W,\boldsymbol{\lambda}) &= \sum_i \{\lambda_i^{out} s_i^{out} + \lambda_i^{in} s_i^{in}\} + \lambda L = \\ &= \sum_i \sum_{j \neq i} \{(\lambda_i^{out} + \lambda_j^{in})\omega_{ij} + \lambda \Theta(\omega_{ij})\} \end{aligned} \tag{15.11}$$

Also in this case the partition function of Eq. (15.8) can be analytically computed and, as a consequence, the probability distribution for the network ensemble is obtained. See (Bargigli 2014) for further information about the model. We indicate this model as Directed Weighted Network model with fixed Strength sequence and Number of Links ($DWNSNL$).

It is well known (Caldarelli et al. 2013) that the bosonic configuration model does not capture the fact that real credit networks are sparse. On the contrary, by imposing in average the number of interbank exposures, we obtain a model describing sparse networks by definition. The main question is how well other real-world

[6] The standard *configuration model* is the network model obtained by imposing the degree sequence rather than the strength sequence.

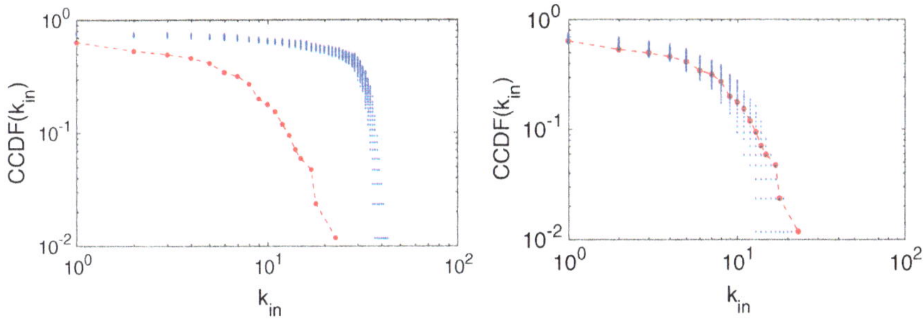

Fig. 15.2 Complementary Cumulative Distribution Function (CCDF) for the real in-degrees (*red dots*) and for the in-degrees of each network in the ensemble (*blue dots*). *Left panel* comparison with bosonic configuration model. *Right panel* comparison with the sparse generalization

characteristics, e.g. the degree distribution, are explained only as a consequence of sparsity.

The results of the fit of e-MID data with these two ensembles are shown in Fig. 15.2. It shows the Complementary Cumulative Density Function (CCDF) of the in-degree distribution[7] of the real network and of the fitted bosonic network without (left) or with (right) the sparsity constraint. The in-degree distribution for the bosonic configuration model is flat before a sharp drop. This is naively due to the fact that the network is fully connected and to the finite number of nodes. On the contrary, the e-MID network (red point) is sparse. Its in-degree distribution is well described by the sparse generalization of bosonic model. We test this result by applying the two-sample Kolmogorov-Smirnov test to the real data and each of the two models. The test rejects the bosonic configuration model but not the sparse generalization.[8]

In order to further investigate the capability of sparse bosonic network to reproduce the topology of the interbank network, in the left top panel of Fig. 15.3 we compare the scatter plot between the real degree and the degree from the simulations of the sparse bosonic model. We observe that large degrees are underestimated by the model, although there is a significant positive correlation between the real sequence and the reconstructed one. Thus the sparse bosonic network shows that the knowledge of the number of links of the network greatly improves the reconstruction when compared with the pure bosonic ensemble.

We compare the bosonic configuration model with the previously introduced method based on relative entropy minimization. In the top right panel of Fig. 15.3 we show the logarithmic difference between the posterior solution of Eq. (15.5) and the mean of the bosonic ensemble as a function of the real weights when these are non vanishing. We notice that for large real weights the two methods agree while a

[7] Similar results are obtained for the out-degree distribution.

[8] We applied the statistical analysis for different aggregation time scales and for different periods in the year 2011 and we obtained always the same results. In this sense, the analyzed properties are stationary for the e-MID market, at least in the year 2011.

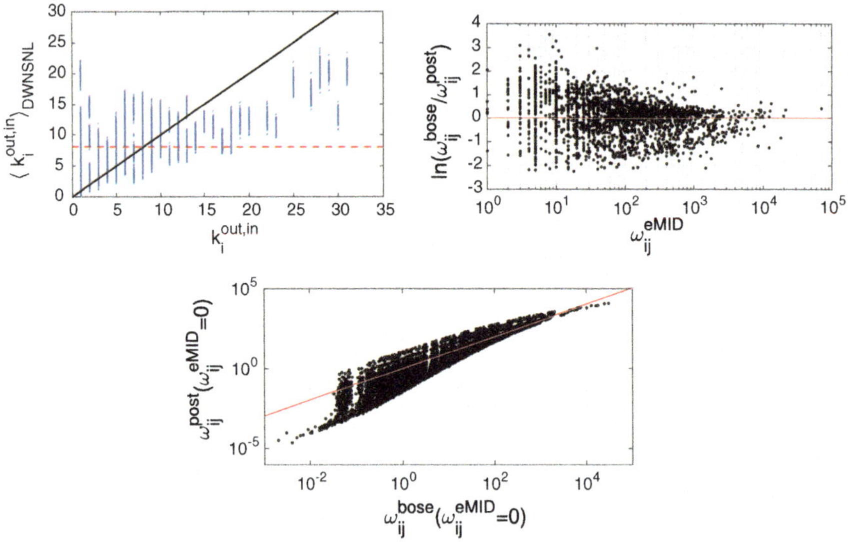

Fig. 15.3 *Left top panel* scatter plot for the real degrees of the aggregate e-MID network in the business period from the 12-20-2011 to the 12-30-2011 and for the average values of the in-degrees on 10 graphs of the sparse bosonic ensemble. The *red dotted line* represents the case of the Erdős-Rényi model according to which each node has in average the same degree set equal to $\frac{L}{N} \approx 8$. *Right top panels* scatter plot of the logarithmic difference of the entries of the weights matrix between the posterior and the mean of the bosonic ensemble as a function of the (non zero) real weights. *Bottom panel* scatter plot of the reconstructed weights with the two methods for zero real weights

significant dispersion is observed for small weights. The bottom panel is the scatter plot of the weights reconstructed by the two methods when the real weights are zero. In this case the correlation is very strong.

15.5 Fitness Model for Network Reconstruction

In this Section we present a third approach for network reconstruction based on the so-called fitness model, firstly introduced in (Caldarelli et al. 2002). The approach with the fitness model addresses all the situations in which there is a strong correlation between the degree and the fitness of a node in the network. The fitness x_i for a node i, x_i^{out} and x_i^{in} in the directed case, can be any non-topological feature determining the probability of creating a link in the network.

In a very general framework, fitnesses are random numbers taken from a given probability distribution $\rho(x)$. An edge from node i to node j is drawn by a Bernoulli trial with success probability equal to $f(x_i^{out}, x_j^{in})$. The *linking* function f is a symmetric function of its arguments and $0 \leq f(x_i^{out}, x_j^{in}) \leq 1$. A fitness model is completely defined once that the function f and the probability distribution ρ are specified. The topological properties of this network model depend on the distribution of fitnesses (or hidden variables) and on the linking function (Boguná and Pastor-Satorras 2003). It has also been proved that by making specific choices for the probability distribution ρ and for the linking function f, the scale-free behaviour appears (Servedio et al. 2004).

The main intuition of the fitness model as a network model is related to the so-called *good-get-richer* mechanism. According to this mechanism, nodes with larger fitness are more likely to become hubs in the network (i.e. to be highly connected).

Different recent works face the problem of network reconstruction via fitness model: (Garlaschelli and Loffredo 2004; Almog et al. 2015) study the case of World Trade Web by associating the fitness with the GDP of a country, (De Masi et al. 2006; Musmeci et al. 2013; Cimini et al. 2015a, b) focus on reconstruction of the e-MID interbank network.

15.5.1 Reconstructing via Fitness Model

According to the original idea of (De Masi et al. 2006), we assume the size of a bank as the fitness of the node representing the bank. In turn, the volume of the interbank exposures represents a measure of the size. Specifically, let us define $x_i^{out} \equiv \frac{(s_i^{out})^*}{\sum_i (s_i^{out})^*}$ and $x_j^{in} \equiv \frac{(s_j^{in})^*}{\sum_j (s_j^{in})^*}$, where with stars we mean the observed values for interbank assets and liabilities. As linking function f, we choose the following, as in (Garlaschelli and Loffredo 2004),

$$f(x_i^{out}, x_j^{in}) = \frac{q x_i^{out} x_j^{in}}{1 + q x_i^{out} x_j^{in}} \tag{15.12}$$

where q is the free parameter of the fitness model.

The reason why the fitness model can be usefully applied for network reconstruction refers to the disassortative mixing shown by credit networks, especially by e-MID (De Masi et al. 2006). Furthermore, as recently highlighted in (Barucca and Lillo 2016), the e-MID interbank market is better described as bipartite instead of showing a core-periphery structure. These two aspects suggest that small borrowers tend to interact to large lenders and viceversa. The linking function f tries to capture this phenomenon.

The expected values for the degrees and the total number of links are simply

$$\sum_{j \neq i} f(x_i^{out}, x_j^{in}) = \langle k_i^{out} \rangle_q, \quad \sum_{i \neq j} f(x_i^{out}, x_j^{in}) = \langle k_j^{in} \rangle_q, \quad \sum_i \sum_{j \neq i} f(x_i^{out}, x_j^{in}) = \langle L \rangle_q. \tag{15.13}$$

Similarly to the sparse bosonic configuration model, we choose the total number of links as known constraint to estimate the free parameter q by last Equation in (15.13).

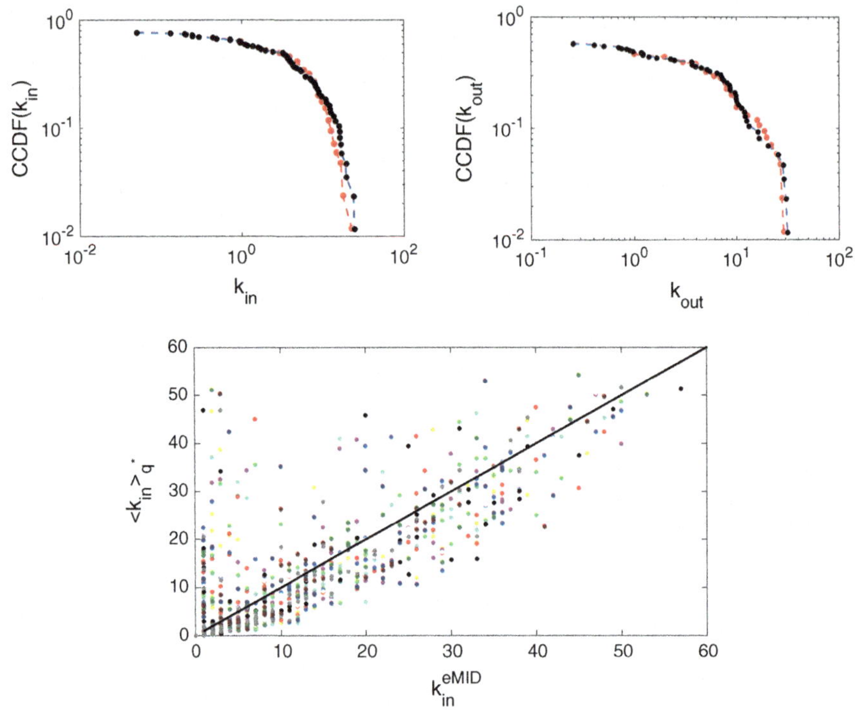

Fig. 15.4 *Top panels* Complementary Cumulative Distribution Function (CCDF) of the real degrees (*red dots*) and of the mean degrees, $\langle k^{in} \rangle_q$ (left) and $\langle k^{out} \rangle_q$ (*right*), obtained by the fitness model (*black dots*). We consider the aggregate e-MID network in the business period from the 12-20-2011 to the 12-30-2011. *Bottom panel* scatter plot of the real in-degree sequence of the monthly aggregate e-MID network for all the months of the year 2011 versus the reconstructed degree sequence $\langle k_{in} \rangle_q$ for each of the considered month. Each *color* is associated with a different month

Once the model is estimated, the linking probability between nodes is fixed and we simply obtain the mean out-(in-)degree of a node as in Eq. (15.13). By reconstructing the mean degree $\langle k^{out(in)} \rangle_q$ for all nodes, we can test, like we did in the ME approach, the null hypothesis that the CCDF of the real degrees and the one of the reconstructed degrees come from the same probability distribution, see the top panels in Fig. 15.4. The two-sample Kolmogorov-Smirnov test does not reject the null hypothesis. As before, the result remains qualitatively the same by changing aggregation time scale or period for the year 2011.

Fitness model well reconstructs the hubs of real world while it does not explain the small connectivity of some large banks,[9] see the bottom panel in Fig. 15.4. However,

[9]In the Italian Overnight Money Market, some links are very persistent in time, that is some banks tend to create credit relations with the same counterparts. By looking at the data aggregates in time, total exposure of a bank may be large but all credit transactions occur with the same counterpart. The persistence of credit relations is not capture by fitness model.

relative errors for network reconstruction are on average smaller than those of the sparse bosonic configuration model.

Once the reconstructed degree sequence is obtained, the main question is how to assign the weight representing the credit exposure between two linked banks.

Consistently with the lack of information beyond the sequence of interbank exposures and the reconstructed degree sequence via fitness model, we can exploit the approach based on maximization of Shannon-Gibbs entropy. In Literature, the model is known as enhanced configuration model (ECM) (Cimini et al. 2015a) and an application to the case of World Trade Web is presented in (Cimini et al. 2015b). The method is totally similar to the one introduced in the previous Section but using as marginals the known strength sequence and the degree sequence reconstructed via fitness model.

The most interesting aspect refers to the capability of the enhanced configuration model in reproducing some of the second-order metrics of the real-world e-MID network. The enhanced configuration model describes quite well disassortative mixing of the real-world credit network, see (Newman 2002, 2003) for definition of these concepts. In our analysis applied to the aggregate e-MID data in the business period from the 12-20-2011 to the 12-30-2011, the ECM estimated on the data, displays an assortative coefficient equal to $-0.19(3)$ while real world shows -0.16 ± 0.11. Moreover the sparse bosonic configuration model does not reproduce the disassortative mixing of real world but it shows a measure that is essentially consistent with zero, 0.01 ± 0.01. Finally, enhanced configuration model is also able to reproduce the clustering coefficient of the e-MID credit network, see (Cimini et al. 2015a).

15.6 Conclusions

In this review, we presented and analyzed three different approaches to the problem of reconstructing a credit network from partial information. We applied our analysis to the data of the Italian overnight interbank market. We show how dense reconstruction methods completely fails in reproducing the topological features of the real world. On the contrary, taking into account that banks have few credit relations is a very important input. In second instance, large banks are more likely to become hubs in the credit network. Fitness model for network formation captures quite well the phenomenon. According to our analysis, reconstructing via fitness model outperforms the other methods when the same input information is used.

Acknowledgements This work is supported by the European Community H2020 Program under the scheme INFRAIA-1- 2014–2015: Research Infrastructures, grant agreement no. 654024 SoBigData: Social Mining & Big Data Ecosystem (http://www.sobigdata.eu).

References

Allen, F. and Gale, D. (2000). Financial contagion. *Journal of political economy*, 108(1):1–33.

Almog, A., Squartini, T., and Garlaschelli, D. (2015). A gdp-driven model for the binary and weighted structure of the international trade network. *New Journal of Physics*, 17(1):013009.

Bacharach, M. (1965). Estimating nonnegative matrices from marginal data. *International Economic Review*, 6(3):294–310.

Bargigli, L. (2014). Statistical ensembles for economic networks. *Journal of Statistical Physics*, 155(4):810–825.

Bargigli, L., di Iasio, G., Infante, L., Lillo, F., and Pierobon, F. (2015). The multiplex structure of interbank networks. *Quantitative Finance*, 15:673–691.

Barucca, P. and Lillo, F. (2015). The organization of the interbank network and how ecb unconventional measures affected the e-mid overnight market. http://arxiv.org/abs/1511.08068.

Barucca, P. and Lillo, F. (2016). Disentangling bipartite and core-periphery structure in financial networks. *Chaos, Solitons & Fractals*, 88:244–253.

Battiston, S., Puliga, M., Kaushik, R., Tasca, P., and Caldarelli, G. (2012). Debtrank: Too central to fail? financial networks, the fed and systemic risk. *Scientific reports*, 2.

Billio, M., Getmansky, M., Lo, A. W., and Pelizzon, L. (2012). Econometric measures of connectedness and systemic risk in the finance and insurance sectors. *Journal of Financial Economics*, 104(3):535–559.

Blien, U. and Graef, F. (1991). Entropy optimization in empirical economic research-the estimation of tables from incomplete information. *JAHRBUCHER FUR NATIONALOKONOMIE UND STATISTIK*, 208(4):399–413.

Boguná, M. and Pastor-Satorras, R. (2003). Class of correlated random networks with hidden variables. *Physical Review E*, 68(3):036112.

Bonanno, G., Caldarelli, G., Lillo, F., and Mantegna, R. N. (2003). Topology of correlation-based minimal spanning trees in real and model markets. *Physical Review E*, 68(4):046130.

Bregman, L. M. (1967). Proof of the convergence of sheleikhovskii's method for a problem with transportation constraints. *USSR Computational Mathematics and Mathematical Physics*, 7(1):191–204.

Caccioli, F., Shrestha, M., Moore, C., and Farmer, J. D. (2014). Stability analysis of financial contagion due to overlapping portfolios. *Journal of Banking & Finance*, 46:233–245.

Caldarelli, G., Capocci, A., De Los Rios, P., and Munoz, M. A. (2002). Scale-free networks from varying vertex intrinsic fitness. *Physical review letters*, 89(25):258702.

Caldarelli, G., Chessa, A., Pammolli, F., Gabrielli, A., and Puliga, M. (2013). Reconstructing a credit network. *Nature Physics*, 9(3):125–126.

Cimini, G., Squartini, T., Gabrielli, A., and Garlaschelli, D. (2015a). Estimating topological properties of weighted networks from limited information. *Physical Review E*, 92(4):040802.

Cimini, G., Squartini, T., Garlaschelli, D., and Gabrielli, A. (2015b). Systemic risk analysis on reconstructed economic and financial networks. *Scientific reports*, 5.

Cont, R., Moussa, A., et al. (2010). Network structure and systemic risk in banking systems. *Edson Bastos e, Network Structure and Systemic Risk in Banking Systems (December 1, 2010)*.

De Masi, G., Iori, G., and Caldarelli, G. (2006). Fitness model for the italian interbank money market. *Physical Review E*, 74(6):066112.

Di Gangi, D., Lillo, F., and Pirino, D. (2015). Assessing systemic risk due to fire sales spillover through maximum entropy network reconstruction. *Available at SSRN 2639178*.

Gai, P. and Kapadia, S. (2010). Contagion in financial networks. In *Proceedings of the Royal Society of London A: Mathematical, Physical and Engineering Sciences*, page rspa20090410. The Royal Society.

Garlaschelli, D. and Loffredo, M. I. (2004). Fitness-dependent topological properties of the world trade web. *Physical review letters*, 93(18):188701.

Garlaschelli, D. and Loffredo, M. I. (2008). Maximum likelihood: Extracting unbiased information from complex networks. *Physical Review E*, 78(1):015101.

Garlaschelli, D. and Loffredo, M. I. (2009). Generalized bose-fermi statistics and structural correlations in weighted networks. *Physical review letters*, 102(3):038701.

Geyer, C. J. and Thompson, E. A. (1992). Constrained monte carlo maximum likelihood for dependent data. *Journal of the Royal Statistical Society. Series B (Methodological)*, pages 657–699.

Iori, G., De Masi, G., Precup, O. V., Gabbi, G., and Caldarelli, G. (2008). A network analysis of the italian overnight money market. *Journal of Economic Dynamics and Control*, 32(1):259–278.

Mastromatteo, I., Zarinelli, E., and Marsili, M. (2012). Reconstruction of financial networks for robust estimation of systemic risk. *Journal of Statistical Mechanics: Theory and Experiment*, 2012(03):P03011.

Mistrulli, P. E. (2011). Assessing financial contagion in the interbank market: Maximum entropy versus observed interbank lending patterns. *Journal of Banking & Finance*, 35(5):1114–1127.

Musmeci, N., Battiston, S., Caldarelli, G., Puliga, M., and Gabrielli, A. (2013). Bootstrapping topological properties and systemic risk of complex networks using the fitness model. *Journal of Statistical Physics*, 151(3-4):720–734.

Newman, M. E. (2002). Assortative mixing in networks. *Physical review letters*, 89(20):208701.

Newman, M. E. (2003). Mixing patterns in networks. *Physical Review E*, 67(2):026126.

Park, J. and Newman, M. E. (2004). Statistical mechanics of networks. *Physical Review E*, 70(6):066117.

Servedio, V. D., Caldarelli, G., and Butta, P. (2004). Vertex intrinsic fitness: How to produce arbitrary scale-free networks. *Physical Review E*, 70(5):056126.

Shannon, C. E. (1948). A mathematical theory of communication. *Bell System Tech. J*, 27:623.

Sheldon, G., Maurer, M., et al. (1998). Interbank lending and systemic risk: an empirical analysis for switzerland. *REVUE SUISSE D ECONOMIE POLITIQUE ET DE STATISTIQUE*, 134:685–704.

Silvestri, L. and Cont, R. (2015). Essays on systemic risk, financial networks and macro-prudential regulation. *PhD Thesis*.

Snijders, T. A. (2002). Markov chain monte carlo estimation of exponential random graph models. *Journal of Social Structure*, 3(2):1–40.

Squartini, T. and Garlaschelli, D. (2011). Analytical maximum-likelihood method to detect patterns in real networks. *New Journal of Physics*, 13(8):083001.

Squartini, T., Mastrandrea, R., and Garlaschelli, D. (2015). Unbiased sampling of network ensembles. *New Journal of Physics*, 17(2):023052.

Upper, C. and Worms, A. (2004). Estimating bilateral exposures in the german interbank market: Is there a danger of contagion? *European Economic Review*, 48(4):827–849.

Van Duijn, M. A., Gile, K. J., and Handcock, M. S. (2009). A framework for the comparison of maximum pseudo-likelihood and maximum likelihood estimation of exponential family random graph models. *Social Networks*, 31(1):52–62.

Wells, S. J. (2004). Financial interlinkages in the united kingdom's interbank market and the risk of contagion.

Chapter 16
Topology of the International Trade Network: Disentangling Size, Asymmetry and Volatility

Anindya S. Chakrabarti

Abstract The international trade network is a complex outcome of numerous purchasing decisions made at the micro level, containing aggregate information about the production technologies and consumers' preferences across countries connected through trade. Thus it acts as a vehicle of spill-over of domestic productivity/preference shocks to the trading partners. The degree is asymmetry in the empirical network indicates different network-wide repercussions of idiosyncratic shocks to individual economies. The structure of the network is shown to be related to the size of the countries and macroeconomic volatility.

16.1 Introduction

The degree of fluctuations of economic entities have been under study for long. High volatility is often representative of risk for example in case of stock prices. Similarly, volatility of GDP of a country represents risk to the consumption decision of the households. GDP is not only the production, but is also the income of the country. The topic of fluctuations in that aggregate income process has received a huge importance in the standard economic literature. However, the sources of such fluctuations are still debated. The candidate explanations range from changing preferences to technological shocks to incorrect expectations. However, after the recent global melt-down that originated in one country and then subsequently propagated to the rest of the world, it is difficult to imagine that the volatility of the countries are purely intrinsic. In fact, given the current degree of openness of the countries in terms of trade flow and capital flow, it seems more realistic that the volatility of the countries are determined jointly, dependent on each other simultaneously.

The international trade network (ITN hereafter) is a rapidly evolving economic entity arising out of millions of small interactions between consumers and producers in different countries (Chaney 2014; Fagiolo et al. 2013; Squirtini and Garlaschelli

---

A.S. Chakrabarti (✉)
Economics area, Indian Institute of Management, Vastrapur, Ahmedabad 380015, India
e-mail: anindyac@iimahd.ernet.in

F. Abergel et al. (eds.), *Econophysics and Sociophysics: Recent Progress and Future Directions*, New Economic Windows,
DOI 10.1007/978-3-319-47705-3_16

2013; Squartini 2011; Squartini et al. 2011). An interesting feature of the ITN is that it shows considerable heterogeneity across its nodes as is evident from Fig. 16.1. Countries vary widely in terms of the degree of strength of their trade relationships with other countries, making some countries substantially more influential in the network than others. During the course of economic globalization, countries have reduced barriers allowing free flow of goods and services across borders which introduces two opposing forces on the partner countries. On one hand, countries have become more vulnerable to the shocks originated in other countries, increasing their own volatility (Easterly et al. 2000). On the other hand, countries can diversify the risks better if they have more trading partners (Haddad et al. 2013). The eventual effect of trade openness is therefore, unclear.

Here, we study the nexus between size, centrality and variance across countries. The relationship between size and variance has been found and studied in the context of firm dynamics (for example, see Riccaboni et al. (2008), Stanley et al. (1996), Amaral et al. (1997)). In case of dynamics of GDP, Canning et al. (1998) showed that size and variance are inversely related. An interesting finding is that the scaling exponent seems to be remarkably close in case of individual business entities and large aggregate entities like countries (Lee et al. 1998). Gabaix (2011) proposed a framework to address this issue. However, our focus is different from the above references in that we consider the network as a whole and show that it is the asymmetry in the trade network that links volatility and size of the countries. Taking this approach further Acemoglu et al. (2012, 2013) showed the possibility of spill-over effects of micro-shocks across the whole economic network. An alternate route (but related to Gabaix (2011)) was taken by di Giovanni and Levchenko (2012) that showed countries with skewed distribution of firm-sizes would show the size-volatility trade-off. However, the effects of the network structure was ignored.

16.2 Properties of the Network and Its Relative Stability

Consider a graph $G = \langle N, A \rangle$ where G is a couplet of the a set of nodes/countries N and the adjacency matrix A that captures the information of the node-to-node connections. In the present case, A_{ij} $(j \neq i \forall i, j)$ is the relative weight of import of country i from j. The diagonal entries are the size of the domestic economies. Hence, the trade network is both *directed* and *weighted* (Fig. 16.1). A noteworthy point is that in the present data set G is found to be fully connected in the sense that each economy is connected to every other economy by imports and exports. Thus the only reasonable choice of studying relative influence would be eigenvector centrality C which is defined as the solution to the equation

$$Ax = x. \tag{16.1}$$

Note that by construction of the adjacency matrix A, it is column-stochastic (see Appendix) with all elements positive. Thus by *Perron-Frobenius theorem* we know

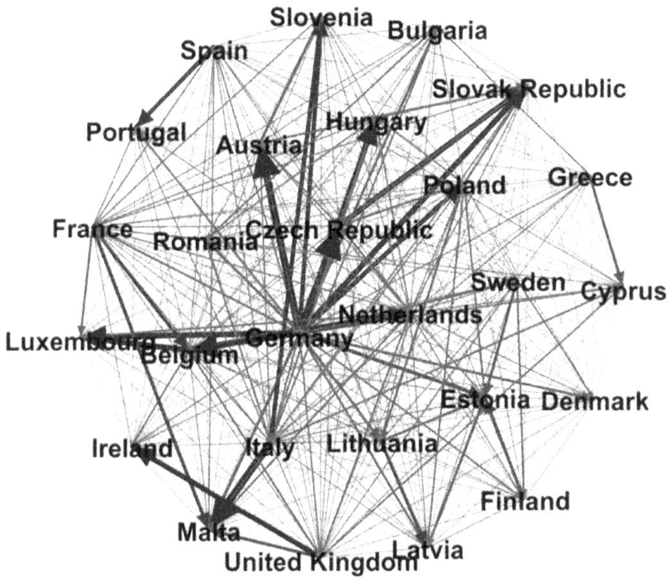

Fig. 16.1 A visual representation of the intra-Europe trade network normalized with respect to the size of the respective economies. The directions of the *arrows* capture the flow of goods and services from source to destination countries with the width indicating the relative strength of the trade connections

that the maximum eigenvalue would be 1. The corresponding eigenvector is taken to be the centrality index (Jackson 2008). This index is self-referential in the sense that if a node is connected to another node with high *prestige*, then the original node's prestige increases. In the current context, the interpretation is if a country has stronger trade relationship with a country which has a well diversified portfolio (and hence lower volatility) then the original country's volatility also decreases because of the connections.

To measure volatility of an economy i, the logged per-capita GDP time series $\{log(Y_{it})\}$ is first decomposed into a trend and a cyclical component with an HP filter,

$$Y_{it} = T_{it}^{Y} + C_{it}^{Y}. \tag{16.2}$$

The second component gives the business cycle fluctuations. The standard deviation of the cyclical component C_{it}^{Y} is taken to be the appropriate measure of macroeconomic volatility. Empirically we find that the volatility of output is negatively related to the centrality of the corresponding country in the trade network. It is important to note that a similar finding has been made in Gray and Potter (2012) albeit with a different interpretation of fluctuations. It considered the volatility of growth rate of GDP in line of Canning et al. (1998) without any particular attention of the cyclical component. Here, I consider detrended series only; hence we are considering business cycle volatility rather than growth volatility.

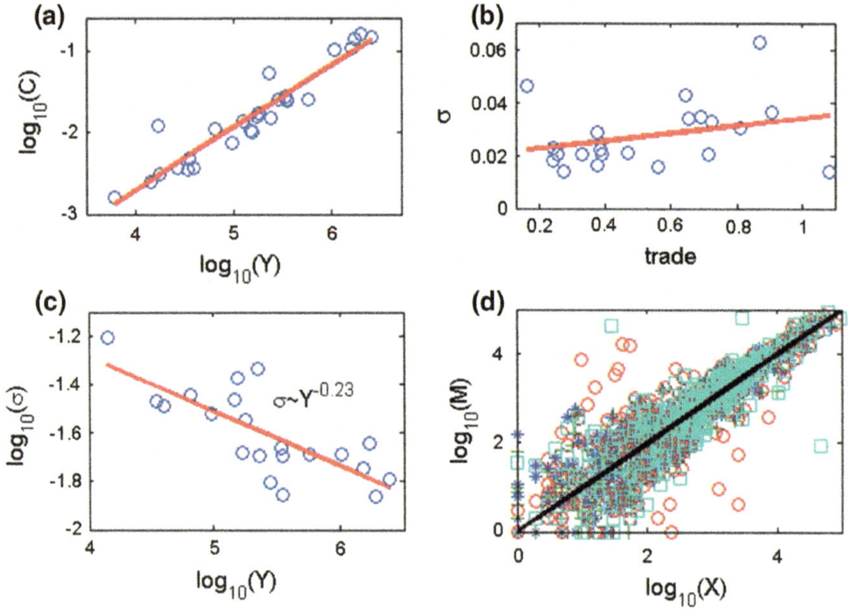

Fig. 16.2 Panel (**a**): Relationship between eigenvector centralities of countries in the international trade network and their corresponding sizes (GDP). Panel (**b**): Relationship between degree of fluctuations and standard trade openness defined as (Export+Import)/GDP. Panel (**c**): A negative relationship between volatility and size. Panel (**d**): Bilateral trade flows between the countries (for years 2001 (+), 2003 (o), 2005 (+), 2007 (□)). It shows that high imports (M) is usually associated with high exports (X) from a given country to another

In the next step, we show that the eigenvector centrality is intimately related to the size of the country. Panel (a) in Fig. 16.2 shows a clear positive relationship between them. The panel (b) shows that volatility mildly increases as the countries become more open in an aggregate sense. This is in accordance with Easterly et al. (2000). However, it is not a reliable indicator of stability as opposed to a more nuanced view that takes into account not only the amount of trade but also the characteristics of the trading partners. Panel (c) shows a negative relationship between size and volatility, as expected (Canning et al. (1998) found a similar pattern; but it used a different metric for volatility). Finally, panel (d) shows that between any two countries high trade flow in one direction is associated with a flow of similar magnitude in the opposite direction. This observation shows that the centrality vector could be derived from the import table (showing the relative strength in the export baskets; as is done here) as well as export table (showing the opposite). Both would give similar results.

As mentioned earlier, the international trade is rapidly expanding both in volume and in scope. In the trade network itself, the ranking of the countries changed appreciably during the period under study. Panel (a) in Fig. 16.3 shows the changes in ranks of the countries. See also panel (c) for the fraction of countries that had different ranks in relative influence. During the whole period the network expanded

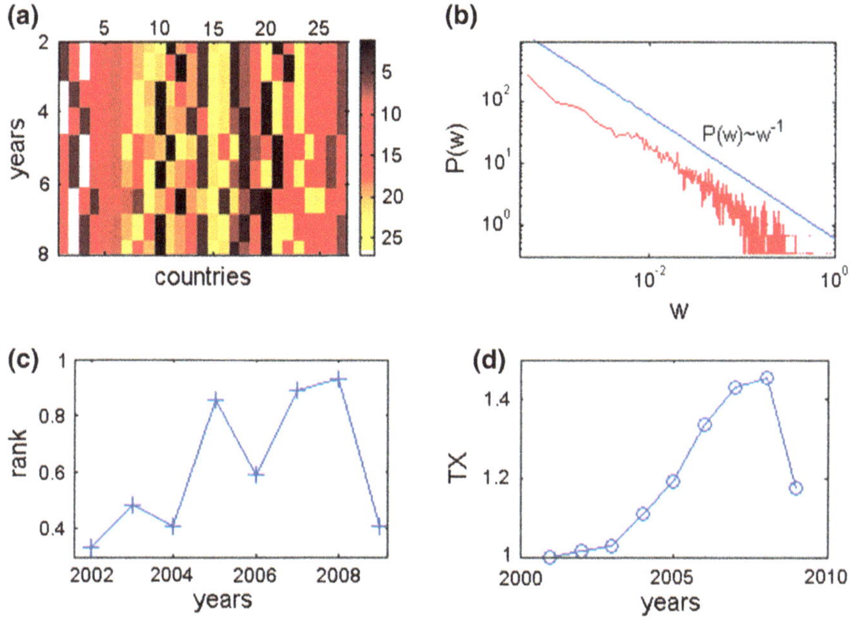

Fig. 16.3 Panel (**a**): A visual representation of changes in relative positions of countries in terms of centrality over the years (2001 to 2009). The colorbar shows the list of countries in alphabetical order. Panel (**b**): Probability distribution function of relative weights (w) assigned by countries to other countries in their respective export baskets. Panel (**c**): The fraction of countries that changed their relative ranking in terms of centrality. Panel (**d**): Evolution of aggregate (intra-Europe) exports relative to the starting year

as a whole as is evident in the increase in volume of trade (panel (d)) relative to the base year. A rapid drop in trade volume is also evident in the after-crash period.

To study the joint evolution of the countries, we also found the cross-correlation matrix of (detrended and logged) quarterly per capita GDP across the countries. The ij-th element of the cross-correlation matrix M^{ρ} is defined as

$$M_{ij}^{\rho} = \frac{\langle (C_i^y - \bar{C}_i^y)(C_j^y - \bar{C}_j^y) \rangle}{\sigma_i \sigma_j}. \tag{16.3}$$

Evidently, the series show substantial correlation in their growth path (see Fig. 16.4). A standard tool to analyze the correlation pattern in multiple time-series in case of finance is eigenvalue decomposition (Plerou et al. 2002) and to use the dominant eigenvalue to find a common global driver of the series (e.g. see Pan and Sinha (2007)). The eigenvalue decomposition shows that the bulk belong to the part generated by random time-series. The existence of outlier shows the interactions between the countries.

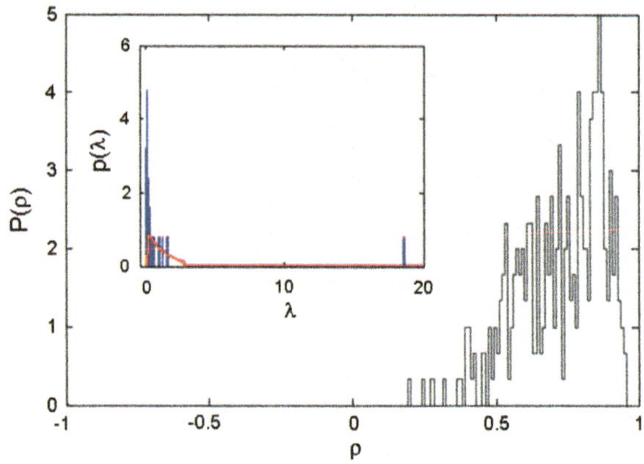

Fig. 16.4 Joint evolution of the GDP series: The probability density function of the correlation coefficients (ρ) between the cyclical components of GDP of the countries. Inset: The distribution of the eigenvalues of the cross-correlation matrix juxtaposed with the corresponding theoretical distribution for a random matrix

16.3 A Prototype Model

Consider the following set of equations describing the evolution of the trade network (with node-specific shocks F_{it}),

$$Y_{it} = \prod_{j}^{N} F_{jt}^{\beta_{ij}} Y_{j,t-1}^{\alpha_{ij}} \tag{16.4}$$

which shows an prototype input-output formalism applied to the trade scenario. Similar approach has been taken to model sectoral interactions (e.g. see Acemoglu et al. (2012)) or firm-firm interaction (see Kelly et al. (2013), Bigio and Lao (2013)). Taking log on both sides, the equation boils down to

$$y_{it} = \sum_{j}^{N} \beta_{ij} f_{jt} + \sum_{j}^{N} \alpha_{ij} y_{j,t-1}, \tag{16.5}$$

where lowercase letter x denotes logarithm of the original variable X. In matrix notation, we can rewrite it as

$$y_t = A.y_{t-1} + B.f_t \tag{16.6}$$

where the matrices $A_{n\times n}$ and $B_{n\times n}$ contains the *linkages* between the countries. Note that the elements of the matrix A captures the strength of the edges existing between countries in terms of flow of goods and services. The second matrix gives the direct, contemporaneous spill-over effects of exogenous shocks. In principle, it captures the effects of common exogenous shocks (say, oil price) or shocks specific to groups of countries (say, a common fiscal policy).

16.3.1 Size-Centrality Relationship

Consider the case $B = [0]_{n\times n}$ or $F_t = [1]_{1\times n}$ i.e. there is no effect of idiosyncratic shocks. Then the dynamical system is convergent, if the A matrix is column-stochastic that is

$$\sum_j A_{ij} = 1. \tag{16.7}$$

However, this is true by construction. When solving for the eigenvector centrality, we used the vector of export weights of each country assigned by the importing country. Since the sum of all weights must add up 1, Eq. 16.7 will be satisfied for all country $i \in N$. This in turn makes the trade-matrix A column-stochastic.

Therefore, Eq. 16.6 effectively becomes equivalent to a Markov chain. From the theory of dynamical systems, it can also be shown that a Markov chain converges to its dominant eigenvector (Chakrabarti 2015). In particular, the solution (if exists) would satisfy

$$y^* = Ay^*. \tag{16.8}$$

Comparing Eqs. 16.1 and 16.8, we see that they are identical with respect to any scaling factor. That is the equilibrium size of the countries is identical to the centrality vector of the same up to any scaling factor since if any y^* is a solution to Eq. 16.8 then so is θy^* for any constant θ. In this proof, the crucial link is provided by the following observation. The equilibrium distribution of a Markov chain is identical to the eigenvector centrality of the network generated by the chain. Therefore we can tie together two different components of the model, one concerning the trade matrix and the other concerning the evolution of size.

16.3.2 Effects of Centrality on Volatility

Solving the basic dynamic equation, we get (assuming $Det(I - A) \neq 0$)

$$y_t = (I - AL)^{-1}B.f_t. \tag{16.9}$$

Given the normalization used above, we cannot assume that $Det(I - A) \neq 0$. However, we can still solve for the growth rate near the steady state and find out volatility of the same. Around the steady state, the growth rate of GDP of the i-th country is

$$
\begin{aligned}
g_{it} &= \frac{y_{it} - y_{i,t-1}}{y_{i,t-1}} \\
&= \frac{y_{it}}{y_{i,t-1}} - 1 \\
&\approx \frac{y_i^* + \sum_j^N \beta_{ij} f_{jt}}{y_i^*} - 1 \\
&= \frac{\sum_j^N \beta_{ij} f_{jt}}{y_i^*}.
\end{aligned} \tag{16.10}
$$

Thus we can express the growth rate as a function of the edge weights, idiosyncratic shocks and the steady state output,

$$
g_{it} \approx \frac{\sum_j^N \beta_{ij} f_{jt}}{y_i^*}. \tag{16.11}
$$

Therefore, the volatility of growth rate

$$
\sigma_i^y \approx \frac{\sqrt{\sum_j^N \beta_{ij}^2 (\sigma_j^f)^2}}{y_i^*}. \tag{16.12}
$$

With $i.i.d.$ shocks on an absolutely symmetric fully connected network $\beta_{ij} = 1/N$, we have

$$
\sigma_i^y \approx \sqrt{N} \left(\frac{\beta \sigma^f}{y_i^*} \right). \tag{16.13}
$$

16.4 Summary

In the current work, we present some evidence in line of Riccaboni et al. (2008) that there exists a robust relationship between size and volatility of countries. The novel feature is that we establish the connection by the missing link that is centrality of a country in the trade network that it is embedded in. We show that following a simple input-output structure we can reconcile the fact that bigger countries are more central. Such economies tend to have more diversified export portfolio allowing them better hedging against idiosyncratic risks (Chakrabarti 2015). Thus they fluctuate less. Hence we establish the connection that bigger economies do fluctuate less, not because of their sheer size but because of the way the trading relationship evolves. In principle, one can provide a demand side explanation featuring the roles played by expansionary monetary/fiscal policies across region and incomplete pass-through of demand shocks through stickyness in nominal variables (Midrigan and Philippon 2011). Chakrabarti (2015) provides a framework to embed full-fledged medium-scale DSGE models in the trade network and contains simulation results showing the match of the model with data. The supply side mechanism presented here in the form of Leontief-type input-output structure provides a complementary description of the system. Further explorations linking the network structure with the macroeconomic factors can be potentially useful.

Acknowledgements Part of this paper is based on the second chapter of my thesis at Boston University (Chakrabarti 2015). I thank Alisdair McKay, Adam Zawadowski, Michael Manove, Bikas K. Chakrabarti, Sitabhra Sinha and Arnab Chatterjee for useful discussions.

Appendix

Description of data

The network is constructed based on the data set for European countries. The reason for choosing European countries is twofold: (1) their relative homogeneity compared to any other countries and (2) availability of data. Since it is known that country-level volatility depends on several country-specific factors like own fiscal/monetary policies, level of development of financial markets (Easterly et al. 2000) apart from global factors like oil-price fluctuations, a relatively homogenous set of countries are chosen which differ in their trade pattern but not in any other dimension (or at least the differences are comparatively small). The yearly time-series data of GDP (1993–2013) is collected from the OECD data base. Country-to-country trade data (complete bilateral flow from 2001 to 2010) is published by Eurostat (2012). For performing the eigenvector decomposition, quarterly GDP data (2000–2012) is used to increase the sample size.

References

D. Acemoglu, V. M. Carvalho, A. Ozdaglar, A. T. Salehi, *The network origins of aggregate economic fluctuations*, Econometrica 80-5, 1977 (2012)

D. Acemoglu, A. Ozdaglar, A. T. Salehi, *The Network Origins of Large Economic Downturns*, unpublished (2013).

R. Albert, A-L Barabasi, *Statistical mechanics of complex networks*, Rev. Mod. Phys. **74** 47 (2002)

L. A. N. Amaral, S. V. Buldyrev, S. Havlin, H. Leschhorn, P. Maass, M. A. Salinger, H. E. Stanley, and M. H. R. Stanley, *Scaling Behavior in Economics: I. Empirical Results for Company Growth*, J. Phys. I France **7**, 621 (1997) and *Scaling Behavior in Economics: II. Modeling of Company Growth*, J. Phys. I France **7**, 635 (1997)

A. Barrat, M. Barthelemy, A. Vespignani, *Dynamical Processes on Complex Networks*, Cambridge Univ. Press, 2008.

S. Bigio, J. Lao, *Financial frictions in production networks*, unpublished (2013)

D. Canning, L.A.N. Amaral, Y. Lee, M. Meyer, H.E. Stanley, *Scaling the volatility of GDP growth rates*, Econ. Lett. **60** 335 (1998)

A. S. Chakrabarti, *Essays on macroeconomic networks, volatility and labor allocation*, Boston University (2015)

T. Chaney, *The network structure of international trade*, Am. Econ. Rev., forthcoming (2014).

J. di Giovanni, A. A. Levchenko. *Country size, international trade, and aggregate uctuations in granular economies*, J. Pol. Econ. **120** 1083, (2012)

W. R. Easterly, R. Islam, J. Stiglitz, *Shaken and stirred: Explaining growth volatility* in Macroeconomic paradigms for less developed countries (2000)

Eurostat, *The trade yearbook*, (2012)

G. Fagiolo, T. Squartini, D. Garlaschelli, *Null models of economic networks: the case of the world trade web*, J. Econ. Inter. & Coord. **8**(1), 75-107, (2013)

X. Gabaix, *The Granular Origins of Aggregate Fluctuations*, Econometrica **79**-3, 733 (2011)

J. Gray, P. B. K. Potter. *Trade and volatility at the core and periphery of the global economy*, Int. Stud. Quart., **56** 793, (2012)

M. Haddad, J. J. Lim, C. Pancaro, C. Saborowski, *Trade openness reduces growth volatility when countries are well diversified*, Canadian J. Econ., **46** 765 (2013)

M. Jackson, *Social and economic networks*, Princeton Univ. Press (2008)

B. Kelly, H. Lustig, S. V. Nieuwerburgh, *Firm volatility in granular networks*, unpublished (2013)

Y. Lee, L. A. N. Amaral, D. Canning, M. Meyer H. E. Stanley, *Universal Features in the Growth Dynamics of Complex Organizations*, Phys. Rev. Lett. **81** 3275 (1998)

V. Midrigan, P. Philippon, *Household leverage and the recession*, unpublished (2011)

R. K. Pan, S. Sinha, *Collective behavior of stock price movements in an emerging market*, Phys. Rev. E **76** 015101-1 (2007)

V. Plerou, P. Gopikrishnan, B. Rosenow, L. A. N. Amaral, T. Guhr, H. E. Stanley, *Random matrix approach to cross correlations in financial data*, Phys. Rev. Lett. **65** 066126 (2002)

M. Riccaboni, F. Pammolli, S. Buldyrev, L. Ponta, E. Stanley,*The size variance relationship of business firm growth rates*, Proc. Natl. Acad. Sci. **105**-50, 19595 (2008)

T. Squartini, G. Fagiolo, D. Garlaschelli (2011) Randomizing world trade. II. A weighted network analysis Phys. Rev. E 84: 046118 (2011)

T. Squartini, G. Fagiolo, D. Garlaschelli (2011) Randomizing world trade. I. A binary network analysis Phys. Rev. E 84: 046117 (2011)

T. Squirtini, D. Garlaschelli, *Economic networks in and out of equilibrium* in Signal-Image Technology & Internet-Based Systems (SITIS), International Conference on pp. 530,537 December (2013)

M. H. R. Stanley, L. A. N. Amaral, S. V. Buldyrev, S. Havlin, H. Leschhorn, P. Maass, M. A. Salinger, H. E. Stanley, *Scaling Behavior in the Growth of Companies*, Nature **379**, 804 (1996)

Chapter 17
Patterns of Linguistic Diffusion in Space and Time: The Case of Mazatec

Jean Léo Léonard, Els Heinsalu, Marco Patriarca, Kiran Sharma and Anirban Chakraborti

Abstract In the framework of complexity theory, which provides a unified framework for natural and social sciences, we study the complex and interesting problem of the internal structure, similarities, and differences between the Mazatec dialects, an endangered Otomanguean language spoken in south-east Mexico. The analysis is based on some databases which are used to compute linguistic distances between the dialects. The results are interpreted in the light of linguistics as well as statistical considerations and used to infer the history of the development of the observed pattern of diversity.

17.1 Introduction

Complexity theory is a major interdisciplinary paradigm which provides a unified framework for natural and social sciences. At an operative level, it is based on a combined application of quantitative and qualitative methods at various phases of research, from observations to modeling and simulation, to the interpretation of com-

J.L. Léonard
Paris-Sorbonne University, STIH 4509 EA, Paris, France
e-mail: leonardjeanleo@gmail.com

E. Heinsalu · M. Patriarca (✉)
NICPB–National Institute of Chemical Physics and Biophysics,
Rävala 10, 10143 Tallinn, Estonia
e-mail: marco.patriarca@kbfi.ee

E. Heinsalu
e-mail: els.heinsalu@kbfi.ee

K. Sharma · A. Chakraborti
SCIS–School of Computational and Integrative Sciences,
Jawaharlal Nehru University, New Delhi 110067, India
e-mail: kiransharma1187@gmail.com

A. Chakraborti
e-mail: anirban@jnu.ac.in

F. Abergel et al. (eds.), *Econophysics and Sociophysics: Recent Progress and Future Directions*, New Economic Windows,
DOI 10.1007/978-3-319-47705-3_17

plex phenomena (Anderson 1972; Ross and Arkin 2009). Among the many applications, ranging from physics to biology and the social sciences, the study of language through the methods of complexity theory has become an attractive and promising field of research. In this contribution we consider the complex and interesting case of the Mazatec dialects, an endangered Otomanguean language spoken in south-east Mexico by about 220,000 speakers (SSDSH 2011–16 2016; Gudschinsky 1955, 1958).

17.1.1 General Method

Language dynamics represents a relevant branch of complexity theory which investigates the classical problems arising in the study of language through novel approaches. Several methods have been imported directly from various scientific disciplines and used to model language from different points of view and at different levels of detail, which complete each other providing, all together, a new informative picture. Among these models and methods, one finds for instance:

(a) simple models addressing the language dynamics of population sizes at a macro- or meso-scopic scale, as in the ecological modeling à la Lotka-Volterra (Heinsalu 2014), which are able to tackle delicate issues such as the perceived status of languages (which directly affect the one-to-one language interaction between individuals) and describe other social features;
(b) nonlinear and stochastic dynamical models, reaction-diffusion equations, etc., which allow one to investigate at a meso-scopic level the most different issues and effects, related, e.g. to population dynamics, the spreading in space of linguistic feature on the underlying physical, economical and political geography (Patriarca and Heinsalu 2009);
(c) individual-based models at the microscopic level, which are used to make numerical experiments to study languages along the perspective of language evolution (Steels 2011) and language competition, i.e., the dynamics of language use in multilingual communities (Solé et al. 2010; Stauffer and Schulze 2005; Wichmann 2008; San Miguel et al. 2005). The latter topic is deeply linked to social interactions, thus the models used have direct connections with social sciences and social dynamics. In fact, linguistic features can be considered as cultural traits of a specific nature and their propagation can be modeled similarly to cultural spreading and opinion dynamics processes (Castellano et al. 2009; San Miguel et al. 2005).

17.1.2 Plan of the Work—Application to Mazatec Dialects

The Mazatec dialects are localized in south-east Mexico. The approximate population of 220,000 speakers is characterized by a highly heterogeneous culture and a locally diversified economic production landscape. The Mazatec dialects have become a classical topic in dialectology, due to the fact that they offer the typical highly complex panorama usually observed when studying cultural landscapes, in particular those characterizing endangered languages (SSDSH 2011–16 2016; Gudschinsky 1955, 1958, 1959; Kirk 1966; Jamieson 1988; Jamieson Carole 1996; Léonard et al. 2012; Léonard and dell'Aquila 2014). This paper consists in the analysis of the Mazatec dialects and in particular their mutual linguistic distances, relying on previous and more recent databases and data analyses by various field-linguists. Such results will be reanalyzed and visualized using the tools of Complex Network Theory, providing us with a measure and a picture of their homogeneity and heterogeneity. Different types of data will be considered, such as those related to the average linguistic Levenshtein distance between dialects (Heeringa and Gooskens 2003; Bolognesi and Heeringa 2002; Beijering et al. 2008) or those extracted by a direct comparison between speakers, i.e., based on the mutual intelligibility of dialects (Kirk 1970; Balev et al. 2016). In Sect. 17.2, relying on the knowledge of the system (and in particular of the values of its main parameters) gained by the work carried out thus far (Kirk's comparative phonological database for interdialectal surveys and fieldwork), we will take into account external constraints such as the ecology of the settlement settings throughout the threefold layered system of Lowlands, Midlands and Highlands, as well as the more recently superposed social and economic impact of postcolonial agro-industrial systems, such as coffee, cattle breeding and sugar-cane (all related, e.g., to the agricultural use of the land). In Sect. 17.3, the comparison between the picture suggested by the complex network analysis of the various data sets (overall sample of lexical categories versus a noun data base, restricted to phonological analysis) and other relevant aspects of the system under study will be carried out. This includes comparison of the linguistic networks with the underlying road networks, physical geography, and economical geography. We will oppose *materiality*, such as ecological settings, to *constructs*, such as dialect areas, to account for the evolution of a very intricate diasystem, ending with a set of proposals for diasystemic geometry as a component of language dynamics as a promising field for Complexity Theory.

17.2 Language Ecology

17.2.1 Ecological Settings

Mazatec has resisted assimilation in the long term, thanks to its demographic weight (more than 200 000 speakers) and to emerging language engineering for literature and education through modern spelling conventions but it is still a very vulnerable language. The data collected in the ALMaz (A Linguistic Atlas of Mazatec; see Léonard et al. 2012) support a pessimistic impression, also considering the collapse of the more recent agrarian systems of coffee crops and cooperatives, the consequences of the Miguel Alemán's dam in the 1950s, still to be seen (see Meneses Moreno 2004; Schwartz and Diana 2016), and a constant drive of migration to urban centres such as Tuxtepec, Tehuacán, Oaxaca, Puebla, México DF, or the USA. The Mazatec area stands in the very centre of the Papaloapam Basin, benefiting from a smooth transition between the plain (e.g., Jalapa de Diaz) and the mountains, West of the Miguel Alemán dam. This ecologically strategic position turned out to be fatal to the Mazatec Lowlands, partly drowned by the Miguel Alemán dam in the mid-50s, when the Rio Tonto, a powerful river connected to the Papaloapam mainstream, was controlled for the benefit of beverage and hydroelectric companies. Sugar cane also demands much water for crops. Patterns of cross-regional integration which had quietly evolved since Olmec times (Killion and Urcid 2001) were disrupted in one of the few regions where native peasants (Mazatec and Chinantec mostly) worked their own microfundio. Maps in Figs. 17.1, 17.2 and 17.3 enumerate the Mazatec municipalities from Baja to Alta Mazateca (Lowlands and Highlands), providing an explicit view of the landscape: to the east, a plain half drowned by the dam (the Lowlands), to the west, a high Sierra mountain chain divided in the south by a canyon—the Cuicatlán Canyon, with the Mazatec small town of Chiquihuitlán, famous for Jamieson's grammar and dictionary, published by the SIL in the late 80s and mid-90s (Jamieson 1988; Jamieson Carole 1996). Figure 17.1 provides an orographic and hydrographic map of the Mazateca area. Figure 17.2 shows the distribution of Municipios over the Mazatec area—the shape of the spots on the maps in

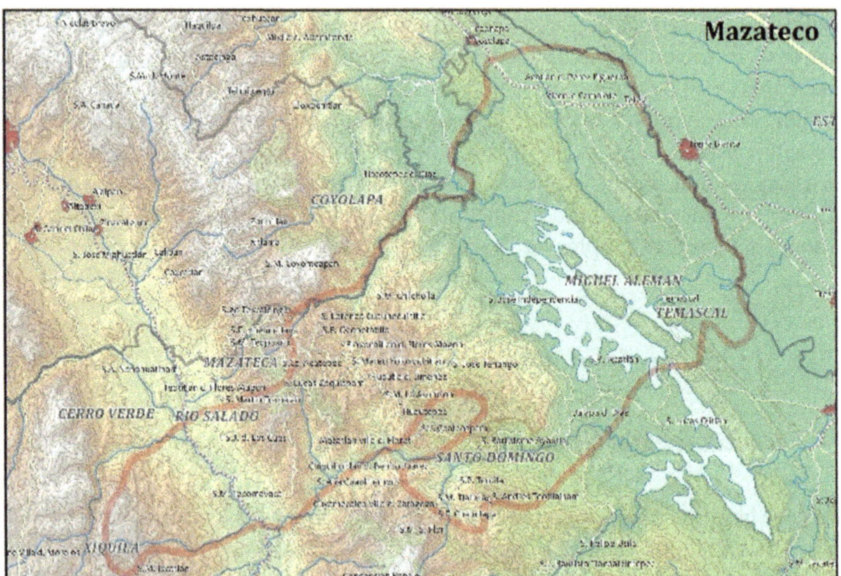

Fig. 17.1 The Mazatec dialect network (localities surveyed in Kirk 1966). Maps: CELE (Vittorio dell'Aquila)

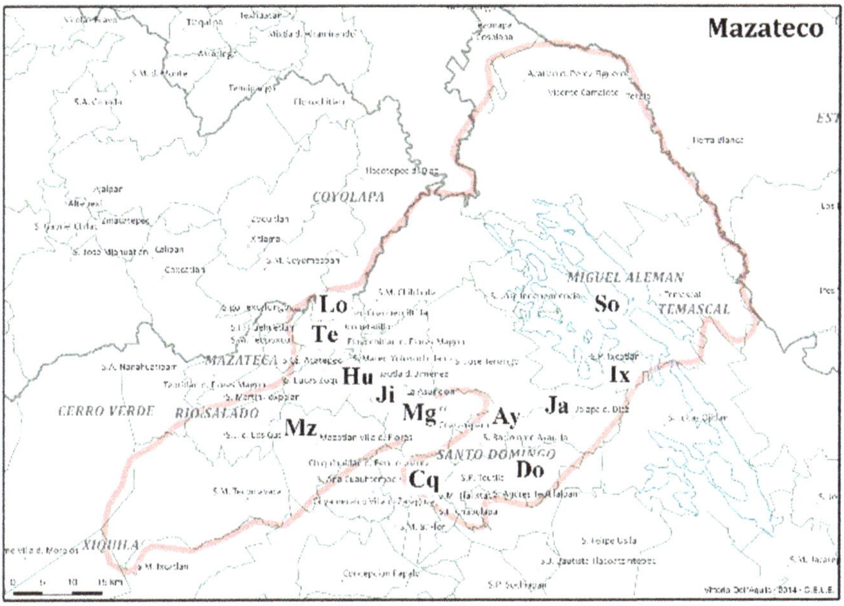

Fig. 17.2 The Mazatec dialect network (localities surveyed in Kirk 1966). Maps: CELE (Vittorio dell'Aquila)

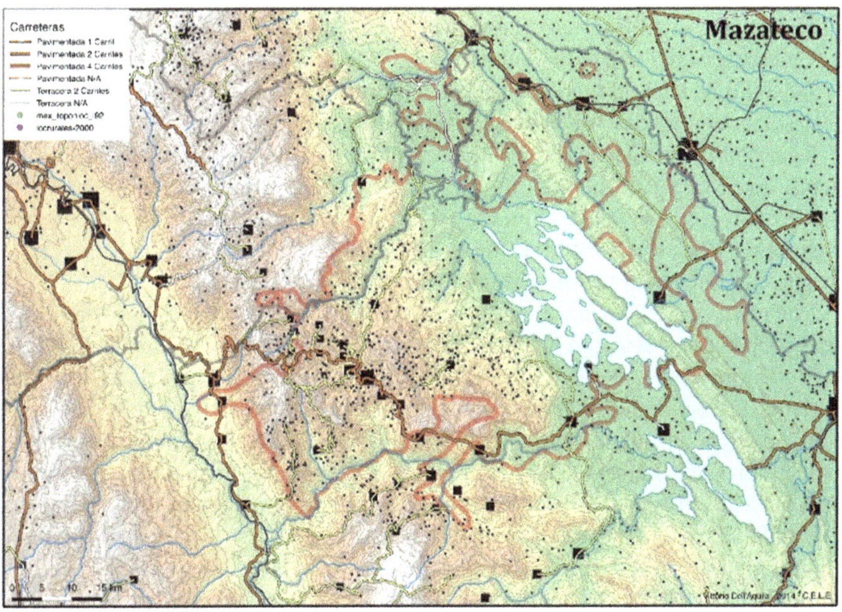

Fig. 17.3 Communal aggregates in the Mazatec area. Map: CELE (Vittorio dell'Aquila). Official census data (2002)

Figs. 17.1 and 17.3 hints at demographic size for each town, whereas Fig. 17.3 points out the *municipios* visited for the ALMaz since 2010 (in this map, only localities already surveyed by Paul Livingston Kirk are mentioned, showing how the ALMaz network is intended to be much larger than in previous dialectological studies, as Kirk 1966; Gudschinsky 1958, 1959). The Mazatec diasystem (Popolocan, Eastern Otomanguean) can be divided into two main zones: the Highlands and the Lowlands. Other subzones can be further distinguished, such as the Midlands (Jalapa de Diaz, Santo Domingo, San Pedro Ixcatlán) within the Lowlands, the Cuicatlán Canyon (Chiquihuitlán) and the Puebla area (see San Lorenzo data below). In short, main dialect subdivisions read as follows, slightly modified from Léonard & Fulcrand 2016:

(1) The Mazatec diasystem: dialects and subdialects

Highland complex: Central Highlands (Huautla de Jiménez, Santa Maria Jiotes, San Miguel Huehuetlán)

Northwestern Highlands: Central Northwestern Highlands (San Pedro Ocopetatillo, San Jeronimo Tecoatl, San Lucas Zoquiapam, Santa Cruz Acatepec, San Antonio Eloxochitlán)

Peripheral Northwestern Highlands (San Lorenzo Cuaunecuiltitla, Santa Ana Ateixtlahuaca, San Francisco Huehuetlán)

Lowland complex: Eastern Lowlands (San Miguel Soyaltepec) Central Lowlands (San Pedro Ixcatlán)

Piedmont or Midlands (Ayautla, San Felipe Jalapa de Diaz, Santo Domingo)

Periphery: South-Western Highlands: Mazatlán Villa de Flores

Cuicatlán Canyon: Chiquihuitlán.

It should be kept in mind that such a classification is not exhaustive but provides only a heuristic framework to observe variation.

The spots on the map in Fig. 17.3 cluster into significant subareas. Behind the dam stands San Miguel Soyaltepec, a very important centre from ancient times, which was probably connected through the plains to the coastal zone of the Papaloapam Basin. From the size of the spots in Fig. 17.3, revealing the demographic weight, we can state that it is still the biggest urban centre in the Mazatec lands.

The town of Acatlán, north of Soyaltepec, is more Spanish speaking than Soyaltepec. Inhabitants of the archipelago inside the artificial lake—within the huge pool created by the dam—use the same variety as in San Miguel Soyaltepec, as do the new settlements, such as Nuevo Pescadito de Abajo Segundo, in the South. A dialect network probably as intricate as that of the North-West Highlands (around San Jernimo Tecoatl) probably existed before the flooding of the microfundio agrarian society of the Lowlands. Most of these dialects merged into mixed dialects, apparently under the strong influence of the Soyaltepec koinè (we use this term as "local speech standard", i.e. pointing at an oral, more than a written koinè, though nowadays a Soyaltepec written koin does exist, strongly supported by local poets and school teachers). This first segment of the Mazatec world makes up the San Miguel Soyaltepec Lowlands segment: a resilient area, with a strong urban constellation going from the newly built Temascal to the industrial town of Tuxtepec, with strong local dialect intercourse and mingling, in a region whose agrarian structure has been drowned by a pharaonic dam project sixty years ago. The consequences of this dramatic redistribution of agrarian resources and property, and of the displacement of over 22 000 peasants, are still to be seen. Linguistically, this event partially enhanced acculturation and assimilation to Spanish under the influence of urban centres such as SM Soyaltepec, but most of all, Temascal, Acatlán, and Tuxtepec. The second area, going from Lowlands to Highlands, covers the western shores of the Miguel Alemán lake, as a twofold stripe, from S. M. Chilchotla and San José Independencia (Midlands) to San Pedro Ixcatlán (Western Lowlands), in the continuity of the plain or the valley, where the important urban centre of Jalapa de Diaz is located. This Midland-Lowland region displays a whole range of small urban centres, dominated by sugar-cane and herding (the agrarian couple *caña y ganado*). Though we should consider Jalapa de Diaz as a subarea of its own, because of its size and its links with other regions, such as the Highlands (Huautla) and the so called *Cañada* or Canyon (Chiquihuitlán and beyond), we may lump both subareas as the Western Plain. The Highlands qualify as the third main area, after the subdivisions of the Lowlands into the SM LL and the Western Plain. In turns, it divides into two subareas: central, with Huautla, and the Western Highlands—a dense network of small urban centres such as San Lucas, San Jernimo Tecoatl, San Lorenzo, San Francisco Huhuetlán, and San Pedro. We will call the fourth complex "the Cañada Connection", where the most conspicuous urban centre is Mazatlán de Flores, on the periphery of the Canyon, and Chiquihuitlán. This is a region of intense language contacts: from Chiquihuitlán downhill through the Canyon, Cuicateco, a Mixtecan language is spoken. Nowadays, the zone seems to have fallen into the hands of the *Narcos*, and the road to Chiquihuitlán is no longer an easy to trip from Jalapa de Diaz, as the ALMaz staff has experienced in recent years. The dialect of a spot such as Santa Maria Tecomavaca, on the western plateau, has scarcely been documented up to now, though it is not so far from neighbouring centres such as Mazatlán or Teotitlán del Camino. Though, it forms a subarea on its own in the Canyon region, because of the low rate of Mazatec speakers as compared to the central area of the Mazatec world, and its location on the plateau, with a tropism outward of the Mazatec area (towards Teotitlán del Camino, Tehuacán, etc.). Strikingly enough, the variety spoken in this peripheral area has more to do with the Northwestern Highlands dialects than with the neighboring Mazatlán area, pointing at strong resettlement dynamics throughout the Mazatec area, far beyond the state of the art knowledge of these phenomena. To us, the main reason lies in the way the coffee economy drained people from the poorest regions of the Midland Outer Belt (Santa Maria Chilchotla, San Mateo Yoloxochitlán), towards the Teotitlán del Camino urban centre, where coffee used to be sold to merchants. Though, the San Juan de los Cùes/Santa Maria Tecomavaca still makes up an original dialect of its own, as several varieties apparently migrated there, from the early 19th to the end of the 20th Century.

The agrarian ecology of these subzones appears in Fig. 17.4. Next, we will deal with sociolinguistic ecology, giving a few hints about linguistic vitality.

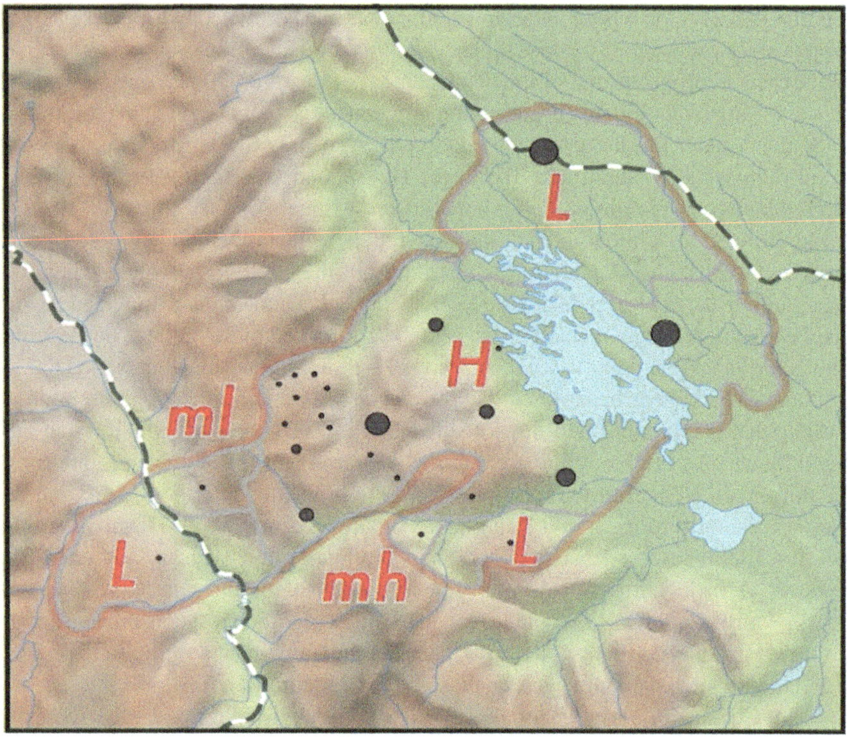

Fig. 17.4 Urban centers and degrees of vitality of Mazatec (Léonard and dell'Aquila 2014)

17.2.2 Sociolinguistics: Vitality Zones

Areas and subareas can also be defined by the sole criterion of the rate of speakers, as in Fig. 17.4: H = High rate of Mazatec speakers (over 75 %), mh = mid-high value, i.e. 50–75 % of the population speaking Mazatec, ml = mid-low density of speakers, i.e. 25–50 %, L = low density, i.e. 0–25 %, in territories considered as traditionally Mazatec. At first sight we can see that the core of the Mazatec area still uses the language intensively (H index), whereas the periphery does not (L on the Eastern shore of the dam and in the Canyon. Two pockets have medium scores: ml at San Juan de los Ces and mh at Chiquihuitlán.

17.3 Dialect Dynamics: A Study in Miniature

The title of this section takes over the subtitle of a seminal paper on Mazatec ethnohistory (Sarah Gudschinsky 1958), in which Gudschinsky claimed that geolinguistics

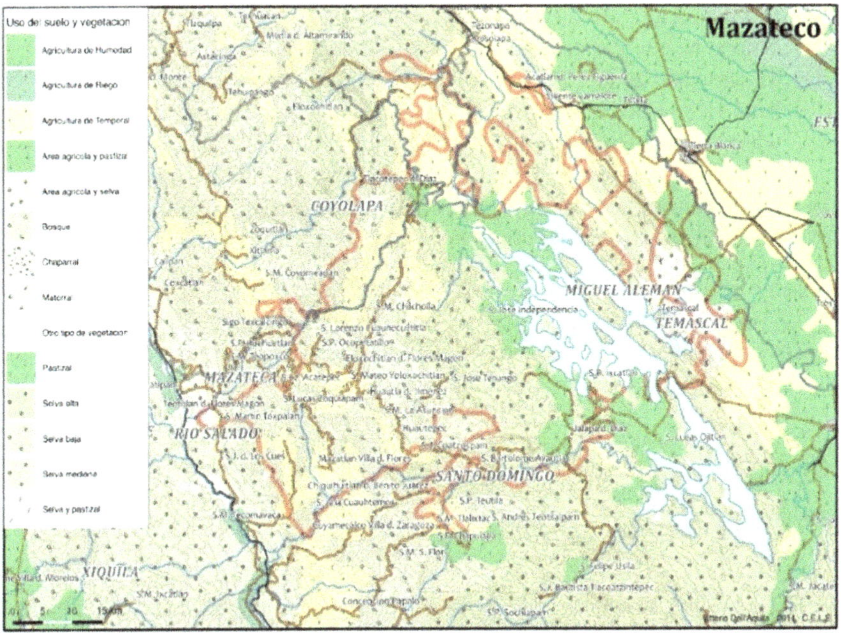

Fig. 17.5 Ecological and agrarian zones in the Mazatec area CELE (Vittorio dell'Aquila)

provided reliable clues to the Proto-Mazatec evolution into five main dialects, through seven periods (Fig. 17.5):

(2) Gudschinsky's 1958 dialect dynamics model:

(A) homogeneity;
(B) slip according to alternating *a and *u;
(C) emergence of a Lowlands dialect, to which Mazatlán (MZ) and San Miguel Huautepec (MG) still belonged—whereas the former is nowadays a peripheral Highlands dialect, the latter strongly clusters with Huautla (HU), in the Central Highlands area;
(D) the Valley dialect emerges (Jalapa, i.e. JA) and differs from MG, then the Southern Valley dialects split from a Northern one, while foreign domination' (Mixtec) takes hold of the region;
(E) the Highlands dialect emerges, and attracts MZ to its circle of influence, roughly during the period 1300 to 1456; two kingdoms compete, in the Highlands and the Lowlands respectively, (F) Western Highlands, MG and Norther Lowlands dialects differ, and Aztec rule takes hold.

A more cautious model without so many details on the Mixtec and Aztec hegemonies was proposed previously by the same author (Gudschinsky 1955) and describes five differentiation periods (or phases):

(3) Gudschinsky's 1955 dialect dynamics model:

(I) Homogeneity, followed by the rise of HU and JA.
(II) Emergence of a transitional buffer zone between HU & JU.
(IIIa) The lowland zone splits in two, with the emerging variety of IX.
(IIIb) Both HU and IX areas diversify: SMt (San Mateo) emerges in the highlands, whereas SO splits from IX. In the buffer zone, MG also emerges. Flows of lexicon and variables still pass from the Lowlands to the Highlands.
(IV) Further and more clear-cut differentiation between IX and SO, in the Lowlands.
(V) Consolidation of the six dialects: sharper frontiers.

In the next section, where the Levenshtein Distance (LD) is applied to Kirk's data on twelve varieties (Kirk 1966) for surveying dialect dynamics, Gudschinsky's models as summarized in (2) and (3) above are very useful to interpret the results and suggest a better overall agreement with Gudschinsky's model (3)—rather than with (2).

17.3.1 Levenshtein Distances

The LD is used to estimate an average linguistic distance between each pair of dialects from the set of the LDs between variants of the same nouns. The LD $L(a, b)$ is a basic measure of the level of difference between two strings a and b, defined as the minimum number of operations (represented by insertions, deletions, or editions) needed to turn a into b or vice versa. For instance, given a = "thia" ("arm", AY) and b = "tşha" ("arm", JI), the Levenshtein distance between these two variants of "arm" is $L(a, b) = 2$, corresponding to the two changes h $\rightarrow$ ş and i $\rightarrow$ h needed to turn one string into the other. The LD $L(a, b)$ has the merit to be simple in definition and use. Its simplicity, however, also represents its limit, due to its independence of the type of the actual operations (whether insertions, deletions, or editions), the number and type of characters changed (e.g. vowels or consonants), and of the order in which they are changed.

We represent two noun variants in dialect i and dialect j of the same semantic meaning, labeled k, as $a_{i,k}$ and $a_{j,k}$. Namely, the locations of dialects are labelled by the index i (or j), running from $i = 1$ ($j = 1$) to the total number of locations $i = NL$ ($j = NL$), while the label k runs over all the $M_{i,j}$ pairs of nouns $a_{i,k}$ and $a_{j,k}$ in dialects i and j with a common semantic meaning, $k = 1, \ldots, M_{i,j}$. For a fixed pair of dialects i and j the corresponding LDs $L^k_{i,j} = L(a_{i,k}, a_{j,k})$ are computed for all the variants k available. The set of LDs thus obtained are then used to compute the average (final) LD $L_{i,j}$ between dialects i and j,

$$L_{i,j} = \frac{1}{M_{i,j}} \sum_{k=1}^{M_{i,j}} L^k_{i,j} \tag{17.1}$$

Notice that this represents a simple arithmetic average, meaning that all the distances are considered to have equivalent statistical weights. Repeating this calculation

	AY	CQ	DO	HU	IX	JA	JI	LO	MG	MZ	SO	TE
AY		0.28	0.20	0.32	0.21	0.24	0.30	0.52	0.29	0.27	0.24	0.29
CQ	0.28		0.30	0.38	0.30	0.33	0.37	0.54	0.34	0.35	0.30	0.34
DO	0.20	0.30		0.33	0.19	0.11	0.33	0.54	0.27	0.26	0.24	0.28
HU	0.32	0.38	0.33		0.32	0.30	0.21	0.53	0.25	0.30	0.24	0.33
IX	0.21	0.30	0.19	0.32		0.22	0.31	0.53	0.29	0.27	0.24	0.25
JA	0.24	0.33	0.11	0.30	0.22		0.32	0.55	0.28	0.28	0.25	0.28
JI	0.30	0.37	0.33	0.21	0.31	0.32		0.55	0.33	0.28	0.24	0.28
LO	0.52	0.54	0.54	0.53	0.53	0.55	0.55		0.55	0.33	0.50	0.50
MG	0.29	0.34	0.27	0.25	0.29	0.28	0.33	0.55		0.25	0.24	0.31
MZ	0.27	0.35	0.26	0.30	0.27	0.28	0.28	0.33	0.25		0.22	0.29
SO	0.24	0.30	0.24	0.24	0.24	0.25	0.24	0.50	0.24	0.22		0.26
TE	0.29	0.34	0.28	0.33	0.25	0.28	0.28	0.50	0.31	0.29	0.26	

Fig. 17.6 A Matrix of LDs for 12 Maztec dialects, 117 cognates. (*source* data from Kirk 1966, data processing: CELE, Vittorio dell'Aquila 2014)

for all pairs of dialects (i, j) allows to construct the "Levenshtein matrix", whose elements are all the average LDs $L_{i,j}$ defined above. The Levenshtein matrix for the twelve Mazatec dialects studied is visualized in the table in Fig. 17.6 (for $NL = 12$ locations, there are $NL(NL - 1)/2 = 60$ such distances).

17.3.2 An Overall Sample for LD

In this section, dialectological data from Kirk (1966) will be measured according to LD (see the chapter on Basque geolinguistics for methodological details). As this algorithm measures and ponders distance between dialects synchronically, most of the results rely upon phonological and morphological patterns. Etyma are not used, contrary to a phylogenetic approach. We will thus consider these results as highlighting ontological distances and complexity between dialects (e.g. the most complex dialect here is LO, in the Poblano area, in the NW outskirts of the Mazatec dialect network.

It is useful to study the network as a function of a threshold T. To this aim, we first normalize all the LDs by dividing them by the largest LD found in the system, so that all the LD values are in the interval (0,1). Then the value $T = 0$ corresponds to prefectly equivalent dialects, while the value $T = 1$ to the farthest couple(s) of dialects. The method consists in setting a threshold on the LDs, i.e., plotting two dialect nodes i and j only if their LD is such that

$$L_{i,j} < T\,, \tag{17.2}$$

At $T = 0$, no link is shown because no dialect is perfectly equal to another dialect. When gradually increasing T, then some dialect nodes become connected producing a linguistic network. At the maximum value $T = 1$ all the dialect nodes appear and are connected to each other. However, not all link strengths are equal. A useful way to plot the network is to make links between nodes thicker if the corresponding LD is smaller, so that they provide an intuitive visual idea of the strength of the linguistic link.

Thus, the threshold $T = 0.20$ shows a choreme (a kernel area, see Goebl 1998: 555). The bolder line uniting JA and DO points at a dialect of its own, whereas the finer line, between DO and IX, resorts to a less organic structural relation, yet rather strong—i.e. a chain, between this basic choreme [JA-DO] with the more autonomous and powerful Lowlands dialect of San Pedro Icxatlán (Fig. 17.7).

With the threshold $T = 0.22$, another choreme shows up, in the Highlands: HU and JI, whereas the inner cohesion within the [IX[DO-JA]] chain is confirmed. This [HU-JI] choreme will soon be connected to the most peripheral dialect, in the Eastern Lowlands (SO), and remains yet unconnected to close neighbors like MG or TE.

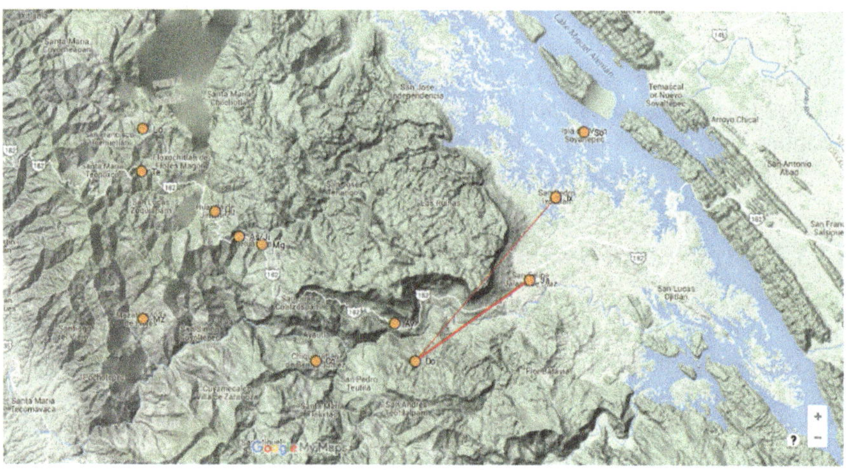

Fig. 17.7 Dialect network with thresholf $T = 0.2$

Fig. 17.8 Dialect network with thresholf $T = 0.22$

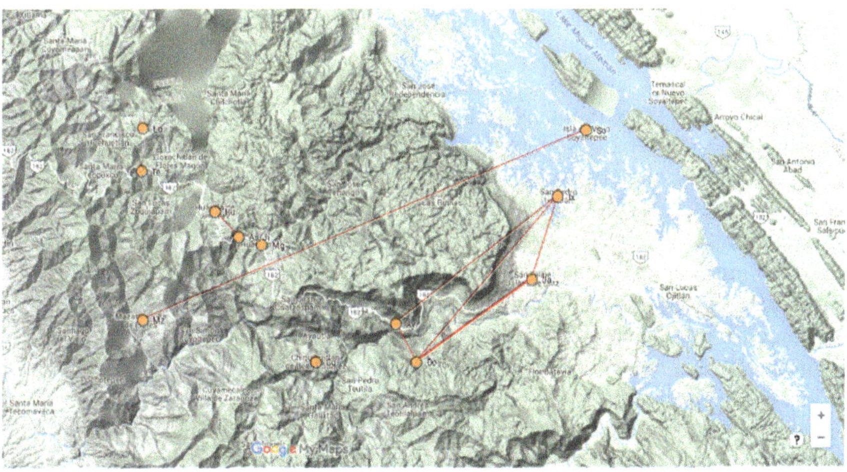

Fig. 17.9 Dialect network with thresholf $T = 0.24$

As we soon shall see, these two choremes now available will soon raise their interconnectivity in the dialect network, enhancing patterns of resilience of a previous feature pool (see Mufwene 2001, 2012, 2013) consistency in the valley (Fig. 17.8).

With the threshold value $T = 0.24$, a complex communal aggregate [[MZ-SO], [HU-JI], [[IX[DO-JA]]] emerges. The pattern now points at two clusters [HU-JI], [[IX[DO-JA]] and one far distant chain [MZ-SO]. As a matter of fact, all these patterns confirm Gudschinsky's model (1955), initially elaborated out of lexicostatistics (Fig. 17.9).

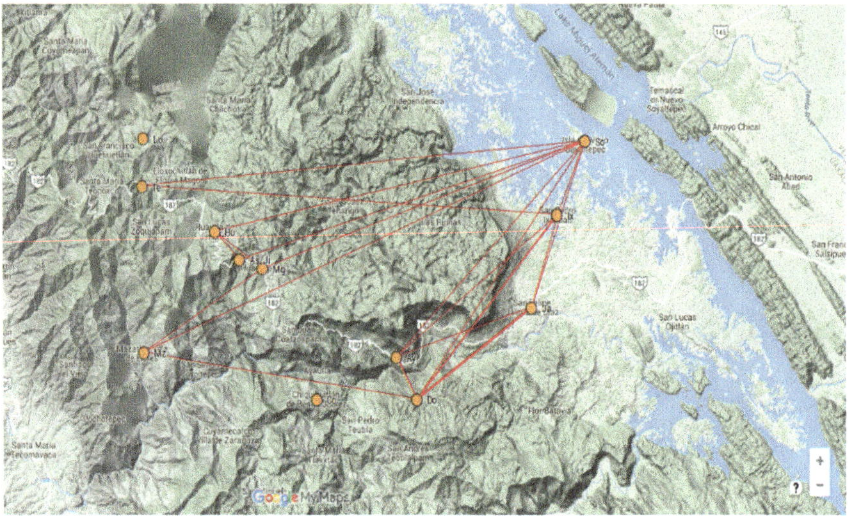

Fig. 17.10 Dialect network with thresholf $T = 0.27$

With T = 0.27, though, the overall picture becomes clearer, and goes far beyond Gudschinsky's expectations, in terms of fine-grained representation of the intricacy of the diasystem; namely, we have a whole complex network with clear-cut communal aggregates: a [TE[SO[IX]] chain, a [HU-JI-MG[SO]] chain, a macro-chain connecting in a most intricate way MZ with the [IX-DO-JA] chain, through AY and MG, working as areal pivots in the Midland and the Highlands respectively. The most peripheral varieties are LO in the Northwestern fringe, and CQ, in the Southwestern border of the Mazatec area. Interestingly enough, these spots are not connected yet in this phase, forming what we can call "default areas" or "default spots", i.e. strongly divergent varieties, which do not correlate tightly enough with the rest of the network to highlight deep geolinguistic structures. Of course, one can cluster these erratic varieties, when elevating the threshold of divergence (Fig. 17.10).

The threshold $T = 0.29$ shows how CQ does correlate with already available clusters—namely, with AY. Nevertheless, AY and CQ strongly differ in all respects, as our own fieldwork gave us evidence recently. The reason why CQ converges somewhat to AY is more due to the transitional status of AY, between the Highlands and the Lowlands, rather than to structural heritage, although indeed, these two variants can be seen as geographical neighbors (Fig. 17.11).

The same could be said of LO, as compared to TE: the former finally connects to the latter in a nearest-neighbor graph, as shown in Fig. 17.12 below (obtained by joining each dialect node only to the one from which it has the shortes LD) although the structural discrepancy is conspicuous. Indeed, LO proceeds from the same historical matrix as TE: the San Antonio Eloxochitlán dialect—not surveyed by Paul Livingston Kirk, but from where we were able to elicit phonological and morphological data in 2011. This nearest-neighbor graph below provides a handy overall

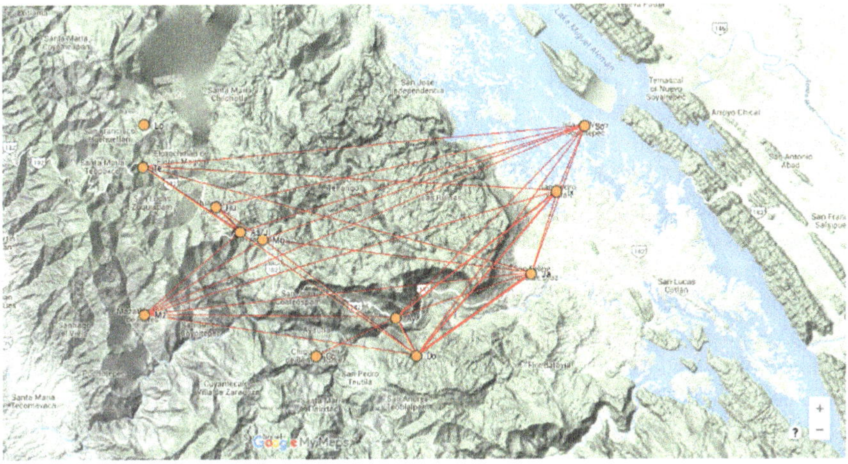

Fig. 17.11 Dialect network with thresholf $T = 0.29$

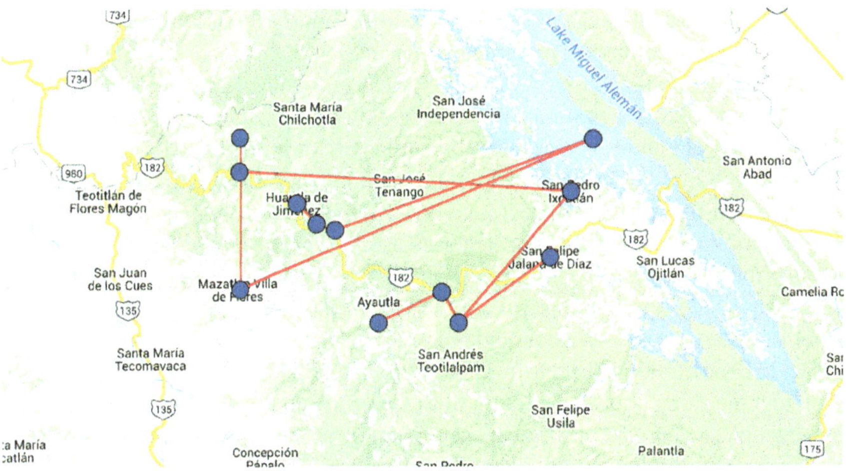

Fig. 17.12 Nearest neighbor network based on the LD distances of 117 cognates, based on the data of (Kirk 1966)

picture of the Mazatec dialect network, on the basis of the LD processing of our 117 cognates: it clearly highlights the far reaching interconnection of Highlands dialects with Lowlands dialects, with macro-chains [TE[IX]], [MZ[SO]] and the intricate cross-areal (i.e. Highlands/Lowlands) cluster [HU-JI-MG[SO]]. Lower range clusters, such as [AY[CQ[DO]]], and choremes, such as [DO-JA] and [HU-JI], as seen previously at stage $T = 0.20$ and 0.22 are also available in this map (Fig. 17.12).

Considering Gudschinsky's model of dialect dynamics (3) above, one can now check to what extent its predictions were right. As a matter of fact, her claim (I) (homogeneity, followed by the rise of Hu and JA) is confirmed by phase $T = 0.22$, which clearly enhances the emergence of two choremes–high and low: [HU-JI] versus [DO-JA].

Gudschinsky's period (II) entails the emergence of a transitional buffer zone between HU & JU. This claim is strongly supported, but also enriched by phases $T = 0.24$ and $T = 0.27$: not only does HU cluster with JI and MG, but AY also clusters with the IX and JA-DO chain. In turn, all these aggregates connect with Lowlands varieties, pointing at the formation of Highlands varieties as a by-product of Lowlands dialect diversification. The ambivalent structural status of MZ, standing far west in the Highlands, though connecting far into the East with SO, and even to IX, through the buffer area of AY, hypothesized by Gudschinsky in both models (2) and (3), is strongly confirmed too. Gudschinsky's Periods (IIIa-b), implying the split of the Lowlands dialect in two (JA vs. IX) on the one hand (IIIa), and on other hand the inner split of the Highlands (i.e. IIIb: HU versus TE, standing for Gudschinsky's SMt, in this dialect network according to Kirk's data) are also confirmed by steps $T = 0.29$ and $T = 0.30$ respectively, as these slots in the graph become more densely interactive with the rest of the dialect network. Though, results here display much more detail on general connectivity than in models in (2) and (3). Last, but not least, period (VI), with further and more clear-cut differentiation between IX and SO, in the Lowlands, is also confirmed by far reaching patterns of connectivity of SO with TE, HU, MZ in the highlands and AY in the Midlands. Results from these 117 cognates (see Léonard 2016: 77–79 for a complete list of items) are not simply congruent with Gudschinsky's hypothesis on dialect dynamics, as summed up in (2) and (3): they provide much more information about the hierarchization and intricacy of differentiation within the Mazatec dialect network. Moreover, they enhance the status and interplay of such (dia)systemic categories as choremes, chains, macro-chains and pivots or buffer zones. They also clearly point at a level of diasystemic organization which supersedes the Stammbaum and the chain level of organization: distant ties, either out of retention, or as an endemic effect of a feature pool (Mufwene 2001, 2012, 2013) of traits inherited from the Lowlands dialects, which carried on mingling together long after the splitting of the main Highlands and Lowlands dialects. For example, many morphological facts point at an inherited stock of inflectional mechanisms in the Lowland dialects and peripheral Northwestern dialects such as LO (in Kirk's data) and San Antonio Eloxochitán (ALMaz data). The link between TE and IX in Fig. 17.12 confirms this trend—whereas the link between HU and SO or MZ and SO may rely more on mere retention, and to an older layer of structural continuity. The sample processed here covered all lexical classes of the Mazatec lexicon, for a set of 117 cognates, from Kirk 1966: verbs, nouns, pronouns, adjectives, adverbs, etc. The results do provide a useful overall picture, but we still suspect this sample to be too heterogeneous, and to blur finer grained patterns of differentiation within the lexicon and grammar. Verbs are especially tricky in Mazatec (Léonard and Kihm 2014; Léonard and Fulcrand 2016) and bias may be induced by elicitation, for instance when the linguist asks for a verb in neutral aspect (equivalent to present tense) and may get an answer in the incompletive (future tense) or completive (past tense), or the progressive aspect, according to pragmatic factors (e.g. verbs such as' die' can hardly be conjugated in the present, as' he dies', and informants are prone to provide completive or incompletive forms, as' he died (recently)' or' he'll (soon) die'). Nouns in Mazatec are far less inflected than verbs—only inalienable nouns, such as body parts and some kinship terms have fusional inflection (see Pike 1948: 103–106). The subset of nouns in the Kirk data base therefore is more likely to provide abundant and much more reliable forms to implement the LD than a sample of all lexical categories.

17.3.3 A Restricted Sample for LD

Although this paper aims at modeling dialect dynamics rather than at providing a description of the language, some data may be useful at this point of the argumentation, in order to get a glimpse at word structure in Mazatec, and related processes on which the LD distance may apply.

All networks emerging from this wider and more consistent sample confirm previous results: at T = 0.45, we find again two choremes—one located in the Southern Lowlands, i.e. [JA-IX], and another located in the Central Highlands, i.e. [HU-JI-MG]. The latter choreme, though makes up a chain with a very interesting dialect, which was already viewed as ambivalent by Gudschinsky: MZ clusters with [HU-JI-MG] in a [MZ[HU-JI-MG]] chain.

The main difference with previous clusters at this stage lays in the boldness of aggregates: MZ would be expected to cluster at a later stage of structural identification with the Highlands choreme, and JA should rather cluster first with DO, instead of telescoping IX. This behavior of the diasystem is due to the lesser complexity of the data, as suggested above when analyzing phonological variables in the table in Fig. 17.13: the simpler the morphological patterns, the more straightforward the results. Bolder chains in Fig. 17.14 give therefore more clear-cut hints at the deep structure of the diasystem. At I = 0.59, an overt extensive rhombus appears, crossing the whole area from west to the east, strongly rooted in MZ in the West and SO in the East, with two lateral extensions: TE in the Northwest and AY in the East. One couldn't dream of a better resume' of most of our previous observations: TE and AY are outstanding actors as pivots, or transitional spots, while MZ, HU and SO had already been noted as crucial innovative dialects, since he early phases of Gudschinsk's models of differentiation—stages (C) and (D) in (2) and stage (IIIa) in (3). At 0.72, a trapezoid resorting more to a parallelogram than to an isosceles shows up, confirming the far reaching links between TE and IX, going all the way down towards AY and CQ to climb up toward MZ and reaching TE in a loop—this geometry actually comprehends the periphery of the diasystem, and may point at a deeper level of structuration.

The Minimum spanning Tree (MST) diagram in Fig. 17.15 endows the Central Highlands dialect JI with enhanced centrality. The fact that the transitional variety of AY in the Midlands is intertwined with another "buffer zone" dialect, according to Gudschinsky's model, confirms details of the deep structure of the dialect network.

#	AY	CQ	DO	HU	IX	JA	JI	LO	MG	MZ	SO	TE
AY	0,000	0,632	0,629	0,668	0,606	0,607	0,636	0,981	0,562	0,573	0,582	0,708
CQ	0,632	0,000	0,717	0,703	0,666	0,704	0,589	0,978	0,627	0,645	0,636	0,688
DO	0,629	0,717	0,000	0,689	0,585	0,334	0,643	1,000	0,608	0,639	0,620	0,703
HU	0,668	0,703	0,689	0,000	0,593	0,655	0,346	0,897	0,402	0,481	0,519	0,550
IX	0,606	0,666	0,585	0,593	0,000	0,599	0,616	0,937	0,574	0,639	0,519	0,586
JA	0,607	0,704	0,334	0,655	0,599	0,000	0,617	0,945	0,594	0,604	0,585	0,675
JI	0,636	0,589	0,643	0,346	0,616	0,617	0,000	0,841	0,377	0,426	0,462	0,502
LO	0,981	0,978	1,000	0,897	0,937	0,945	0,841	0,000	0,883	0,892	0,884	0,870
MG	0,562	0,627	0,608	0,402	0,574	0,594	0,377	0,883	0,000	0,446	0,490	0,539
MZ	0,573	0,645	0,639	0,481	0,639	0,604	0,426	0,892	0,446	0,000	0,511	0,567
SO	0,582	0,636	0,620	0,519	0,519	0,585	0,462	0,884	0,490	0,511	0,000	0,574
TE	0,708	0,688	0,703	0,550	0,586	0,675	0,502	0,870	0,539	0,567	0,574	0,000

Fig. 17.13 LD data from Kirk 1966: 311 nouns

A minimum spanning tree is a spanning tree of a connected, undirected graph such that all the N (here $N = 12$) dialects are connected together with the minimal total weighting for its $N - 1$ edges (total distance is minimum). The distance matrix defined by the LDs among the dialects was used as an input to the inbuilt MST function in MATLAB (See Matlab documentation for details). Here we state Kruskal and Prim algorithms for the sake of completeness of the present article.

Description of the two algorithms:

- Kruskal—*This algorithm extends the minimum spanning tree by one edge at every discrete time interval by finding an edge which links two separate trees in a spreading forest of growing minimum spanning trees.*
- Prim—*This algorithm extends the minimum spanning tree by one edge at every discrete time interval by adding a minimal edge which links a node in the growing minimum spanning tree with one other remaining node.*

  Here, we have used Prim's algorithm to generate a minimum spanning tree.

The dendrogram in Fig. 17.16 does not only provide an overall picture of the dialect network: it tells us more about the intricacy of communal aggregates and layers of differentiation. It also solves a few problems raised by discrepancies between model (2) and (3) and our results. In this *Stammbaum*, Highlands dialects actually cluster with Lowlands dialects, while Southern Midlands dialects cluster together with a "default" variety—CQ, a near neighbor in the South. In the inner cluster of the

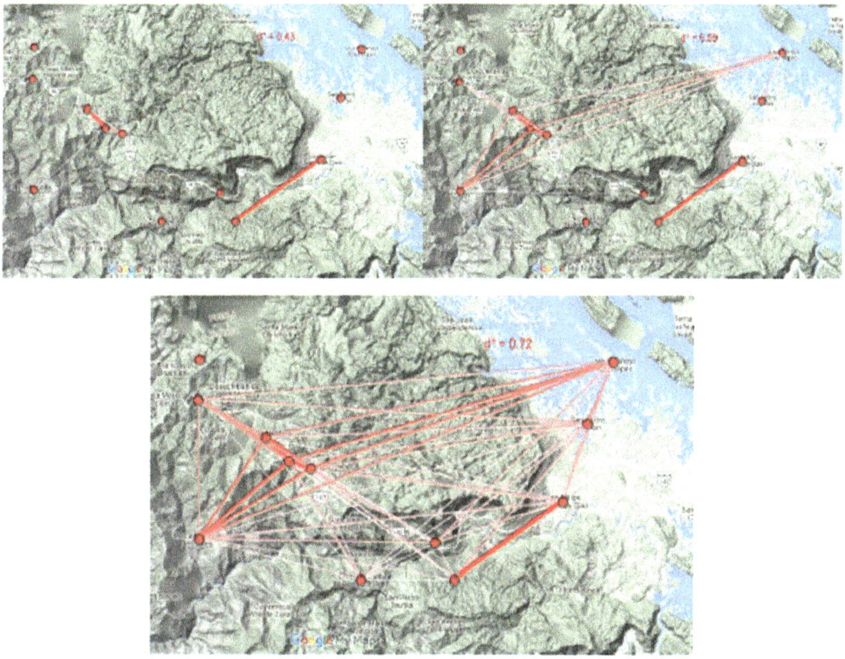

Fig. 17.14 LD applied to nouns in Kirk's data. Three thresholds of normalized mean distance

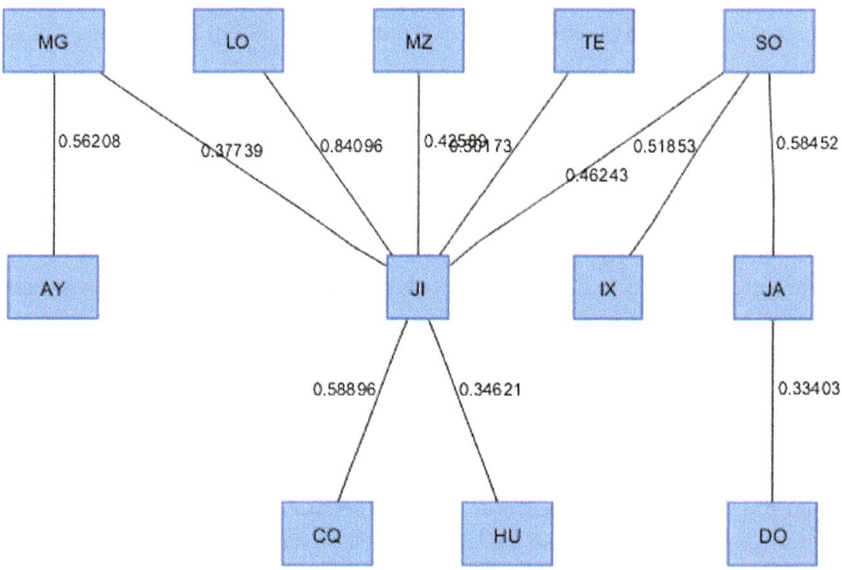

Fig. 17.15 Minimum spanning tree based on the LD applied to nouns in Kirk's data

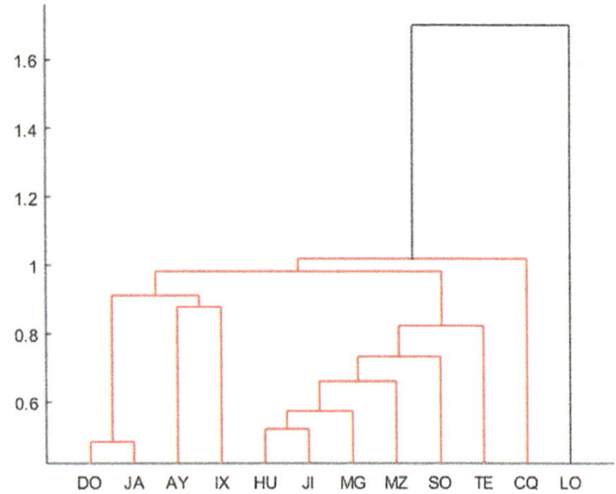

Fig. 17.16 LD applied to nouns in Kirk's data: Dendrogram

dendrogram including Highlands dialects, we come across the [MZ[HU-JI-MG]] chain we are already familiar with, on the one hand, and, on the other hand, a quite heterogeneous subcluster made up of a [IX-SO] chain, associated to the far distant TE Northwestern Highlands dialect, usually classified within the Highland dialects. Last, but not least, the LO dialect, though we can consider it as a byproduct of a recent Northwestern dialect overdifferentiation (i.e. from TE), stands on its own, as if it would classify as a totally different language—which it is not, although its differences are indeed phonologically conspicuous, because of recent vowel shifts i → e, e → a, a → o, u → ï.

A dendrogram is basically a tree diagram. This is often used to depict the arrangement of multiple nodes through hierarchical clustering. We have used the inbuilt function in MATLAB (see MATLAB documentation) to generate the hierarchical binary cluster tree (dendrogram) of 12 dialects connected by many U-shaped lines (as shown in Fig. 17.16), such that the height of each U represents the distance (given by LD) between the two dialects being connected. Thus, the vertical axis of the tree captures the similarity between different clusters whereas the horizontal axis represents the identity of the objects and clusters. Each joining (fusion) of two clusters is represented on the graph by the splitting of a vertical line into two vertical lines. The vertical position of the split, shown by the short horizontal bar, gives the distance (similarity) between the two clusters. We set the property "Linkage Type"as "Ward's Minimum Variance", which requires the Distance Method to be Euclidean which results in group formation such that the pooled within-group sum of squares would be minimized. In other words, at every iteration, two clusters in the tree are connected such that it results in the least possible increment in the relevant quantity, i.e., pooled within-group sum of squares.

In spite of these discrepancies with expected taxon, the main lesson of this dendrogram lays in the tripartition [Midlands[Highlands-Lowlands]], and the confirmation of the [MZ[HU-JI-MG]] chain. In Fig. 17.17, the two-dimensional projection from

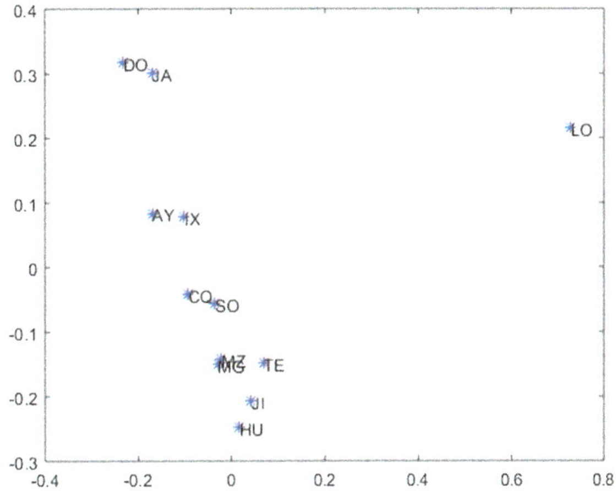

Fig. 17.17 Two-dimensional projection from multi-dimensional scaling analysis (in linguistic space). Nouns in Kirk's data

Multi-Dimensional Scaling (MDS) analysis mends up the formal oddities we already mentioned, i.e. TE clustering so far from HU, and CQ so close to AY. This representation, obtained with the same data, is far more congruent with standard taxonomy of Mazatec dialects, as in (1) above: it displays a constellation of choremes as [DO-JA] and [JI-HU], and more loosely tightened chains such as [AY[IX]], [MZ[MG[TE]]] and a fairly distant chain [CQ[SO]]. LO, again, stands far apart, as a strongly innovative dialect as far as phonology is concerned—with strong consequences on morphology too.

MDS is a method to analyze large scale data that displays the structure of similarity in terms of distances, obtained using the LD algorithm, as a geometrical picture or map, where each dialect corresponds to a set of coordinates in a multidimensional space. MDS arranges different dialects in this space according to the strength of the pairwise distances between dialects—two similar dialects are represented by two set of coordinates that are close to each other, and two dialects behaving differently are placed far apart in space (see Borg 2005). We construct a distance matrix consisting of $N \times N$ entries from the N time series available, defined the using LD. Given D, the aim of MDS is to generate N vectors $x_1, \ldots, x_N \in \Re^D$, such that

$$\|x_i - x_j\| \approx d_{ij} \qquad \forall i, j \in N, \tag{17.3}$$

where $\|.\|$ represents vector norm. We can use the Euclidean distance metric as is done in the classical MDS. Effectively, through MDS we try to find a mathematical embedding of the N objects into $\Re^D$ by preserving distances. In general, we choose the embedding dimension D to be 2, so that we are able to plot the vectors x_i in the form of a map representing N dialects. It may be noted that x_i are not necessarily unique under the assumption of the Euclidean metric, as we can arbitrarily translate and rotate them, as long as such transformations leave the distances $\|x_i - x_j\|$ unaffected. Generally, MDS can be obtained through an optimization procedure, where $(x_1, \ldots, x_N)$ is the solution of the problem of minimization of a cost function, such as

$$\min_{x_1,\ldots,x_N} \sum_{i<j} (\|x_i - x_j\| - d_{ij})^2. \tag{17.4}$$

In order to capture the similarity among the dialects visually, we have generated the MDS plot of 12 dialects. As before, using the International Phonetic Alphabets from the database as an input, we computed the distance matrix using the LD algorithm. The distance matrix was then used as an input to the inbuilt MDS function in MATLAB. The output of the MDS were the sets of coordinates, which were plotted as the MDS map as shown in Fig. 17.17. The coordinates are plotted in a manner such that the centroid of the map coincides with the origin (0, 0).

17.4 Conclusion and Prospects

As Nicolaï and Ploog put it (Nicolaï and Ploog 2013: 278), one has to consider two types of categories, when tackling anything which looks like—or is supposed to work as—frontiers: on the one hand, *matter* or materiality, on the other hand *constructs*. *Matters* or materialities rank as follows: geography, geology, biology, ecology, and they partly shape the world we live in, as we are indeed a very adaptive species. *Constructs*, instead, should be clearly divided in two: *compelling patterns* on the one hand, *elaborations* on the other hand. The former range from social constraints or norms, laws, beliefs and habits to economic systems; the latter from models to reforms, according to the activities developed in communal aggregates, in reaction to the environment and its contradictions.

In this case, *matters* do matter a lot, as the Mazatec diasystem is vertically structured, from the Lowlands to the Highlands, and some bigger and older centers or town dialects, as JA, HU, MZ, IX indeed weight more than mere villages or hamlets (as JI, MG, AY, CQ, LO). The fact that SO was so peripheral, and ended up as a village nested on the top of a resilient hill above the Miguel Aleman dam, as the village called Viejo Soyaltepec, has consequences on the evolution of certain components of the Mazatec diasystem. The intrusion and the violent reshaping of the whole ecological and socioeconomic settings since the end of the XIXth century, though mercantile activities, instead, have resorted to elaborative constructs, and these have played a strong role too, in smashing previous *compelling patterns* of intercommunal solidarity or, on the contrary, enmity. *Matter* and materialities constantly change in nature, indeed, as biology and geology teach us. But cultural *constructs* change even faster, and they may even loop, recede and proceed, in a nonlinear way—as do diasystems throughout history, and so does the Mazatec diasystem in the first place.

But the higher plateau or level in the realm of constructivism and elaboration has to be sought in our models and methods to gather and proceed data, as we did here, handling Kirk's cognate sets, initially collected in order to make a sketch of comparative phonology. We turned it into something quite unexpected, as alchemists used to dream of turning stones or dust into gold. We saw how quantitative tools designed to measure dialect distance, as the Levenshtein algorithm, can provide clues from a Complexity Theory standpoint. Various data sets and a variegated array of computational methods (multilayered normalized means, minimum spanning tree, multi-dimensional scaling analysis, etc.) applied on these raw sets of data opened the way to a labyrinth of constructs and representations, which teach us a lot about what mattered, in the past, and what matters and will, today and for the future, in such a strongly diversified communal aggregates that makeup the Mazatec *small world* (Léonard and dell'Aquila 2014).

A world full of complexity, whose survey with the help of Complexity Theory methods suggest that tree-models *(Stammbaum)*, chain models, choremes and buffer zones or transitional areas are not sufficient to grasp geolinguistic complexity. We also have to resort to concepts as pivots, default varieties, and a few more. Neither is the punctuated equilibrium (Dixon 1997) concept enough, as the Mazatec dialect network geometry shows an intricate web of constant interactions. The valley leading from the Lowlands to the Highlands has not only once in a while served as a bottleneck: it seems to be a highway for diffusion and linguistic change which never rests. Corridors from the Northern Midlands, as Santa Maria Chilchotla, and the San José enango area, between HU and San José Independencia, may also account for this multisource and multidirectional percolation of change and metatypes between communal aggregates. The intricate geometry of diasystems has still to be disentangled, and this Mazatec case study provides but a glimpse at how to tackle this issue. Complexity Theory undoubtedly should be at the forefront of such a crucial endeavor, for the understanding of how complex adaptive and cooperative systems such as language and society work and mingle together.

17.5 Abbreviations

AY = Ayautla, CQ = Chiquihuitlán, DO = Santo Domingo, IX = San Pedro Ixcatlán, JI = Jiotes (or, HU = Huautla, JA = Jalapa, LO = San Lorenzo, MG = San Miguel Huautla, SMt = San Mateo Yoloxochitlán, SO = San Miguel Soyaltepec, TE = San Jernimo Tecoatl (abbreviations as in Kirk 1966).

Acknowledgements M.P. and E.H. acknowledge support from the Institutional Research Funding IUT (IUT39-1) of the Estonian Ministry of Education and Research. K.S. thanks the University Grants Commission (Ministry of Human Research Development, Govt. of India) for her junior research fellowship. A.C. acknowledges financial support from grant number BT/BI/03/004/2003(C) of Government of India, Ministry of Science and Technology, Department of Biotechnology, Bioinformatics Division and University of Potential Excellence-II grant (Project ID-47) of the Jawaharlal Nehru University, New Delhi, India.

References

Anderson, P.W. 1972. More Is Different, Science 177, 393–396.

Balev Stefan, Jean Léo Léonard & Gérard Duchamp 2016. "Competing models for Mazatec Dialect Intelligibility Networks", in Léonard, Jean Léo; Didier Demolin & Karla Janiré Avilés Gonzlez (eds.). 2016. Proceedings of the International Workshop on Structural Complexity in Natural Language(s) (SCNL). Paris, 30–31 May 2016: Paris 3 University - Labex EFL (PPC11). Available on http://axe7.labex-efl.org/node/353.

Beijering, K, C. Gooskens & W. Heeringa 2008. "Predicting intelligibility and perceived linguistic distance by means of the Levenshtein algorithm", Amsterdam, Linguistics in the Netherlands, 2008), p. 13–24.

Bolognesi, R. & W. Heeringa 2002. "De invloed van dominante talen op het lexicon en de fonologie van Sardische dialecten". In: D. Bakker, T. Sanders, R. Schoonen and Per van der Wijst (eds.). Gramma/TTT: tijdschrift voor taalwetenschap. Nijmegen University Press, Nijmegen, 9 (1), p. 45–84.

Borg, I. and Groenen, P., Modern Multidimensional Scaling: theory and applications (Springer Verlag, New York, 2005).

Castellano C., S. Fortunato, V. Loreto, Statistical physics of social dynamics, Rev. Mod. Phys. 81 (2009) 591.

Dixon, Robert, M. W. 1997. The Rise and Fall of Languages, Cambridge, Cambridge University Press.

Goebl, Hans 1998. "On the nature of tension in dialectal networks. A proposal for interdisciplinary research", in Altmann, Gabriel & Walter Koch (eds.) Systems. New Paradigms for the Human Sciences, Berlin, Walter de Gruyter: 549–571.

Gudschinsky, Sarah, 1955, "Lexico-Statistical Skewing from Dialect Borrowing", IJAL 21(2), 138–149.

Gudschinsky Sarah, "Mazatec dialect history", Language, n 34, 1958, p. 469–481.

Gudschinsky Sarah, 1959. Proto-Popotecan. A Comparative Study of Popolocan and Mixtecan, IJAL, n 25-2.

Heeringa, W. & C. Gooskens 2003. "Norwegian dialects examined perceptually and acoustically", Computers and the Humanities, 57 3: 293–315.

Heinsalu E., Marco Patriarca, Jean Léo Léonard, 2014.The role of bilinguals in language competition, Adv. Complex Syst. 17, 1450003.

Jamieson Carole, 1996. Diccionario mazateco de Chiquihuitlán, Tucson, SIL.

Jamieson Carole, 1988. Gramática mazateca. Mazateco de Chuiquihuitlán de Juárez, México D.F, SIL.

Killion Thomas & Javier Urcid 2001. "The Olmec Legacy: Cultural Continuity and Change in Mexico's Southern Gulf Coast Lowlands", Journal of Field Archaeology, 28 1/2: 3–25

Kirk, Paul Livingston 1966. Proto-Mazatec phonology. PhD dissertation, University of Washington.

Kirk, Paul Livingston 1970. "Dialect Intelligibility Testing: The Mazatec Study", International Journal of American Linguistics, Vol. 36, 3: 205–211.

Léonard, Jean Léo ; Vittorio dell'Aquila & Antonella Gaillard-Corvaglia 2012. "The ALMaz (Atlas Lingstico Mazateco): from geolinguistic data processing to typological traits", STUF, Akademie Verlag, 65-1, 78–94.

Léonard, Jean Léo & Alain Kihm, 2014, "Mazatec verb inflection: A revisiting of Pike (1948) and a comparison of six dialects", Patterns in Mesoamerican Morphology, Paris, Michel Houdiard Editeur, p. 26–76.

Léonard, Jean Léo & dell'Aquila, Vittorio 2014. "Mazatec (Popolocan, Eastern Otomanguean) as a Multiplex Sociolinguistic Small World"', in Urmas Bereczki (ed.). The Languages of Smaller Populations: Risks and Possibilities. Lectures from the Tallinn Conference, 1617 March, 2012, Tallinn, Ungarian Institute's Series: Miscellanea Hungarica: 27-55.

Léonard Jean Léo 2016. "Diversification, Diffusion, Contact: Modélisation géolinguistique et complexité". Lalies, 36, E.N.S. de Paris: 9–79.

Léonard Jean Léo & Julien Fulcrand 2016. "Tonal Inflection and dialectal variation in Mazatec", in Palancar, Enrique & Jean Léo Léonard (eds), in Palancar, E. & Léonard, J. L. (eds.) Tone & Inflection : New Facts and New perspectives, Trends in Linguistics. Studies and Monographs, 296, Mouton de Gruyter: 165–195

Meneses Moreno, Ana Bella 2004. Impacto político, social y cultural de la presa Miguel Alemán en la comunidad mazateca de la isla del viejo soyaltepec, Master Thesis, Mexico, Universidad Autonoma Metropolitana (UAM).

Mufwene, Salikoko S., 2001. The ecology of language evolution, Cambridge, Cambridge University Press.

Mufwene, Salikoko, 2012. Complexity perspectives on language, communication, and society, in Ángels Massip-Bonet & Albert Bastardas-Boada, Springer Verlag: 197–218.

Mufwene, Salikoko S., 2013. "The ecology of language: some evolutionary perspectives", in Elza Kioko Nakayama Nenoki do Couto & al. Da fonologia á ecolingustica. Ensaios em homenajem a Hildo Honrio do Couto, Brasilia, Thesaurus, pp. 302–327.

Nicolaï, Robert & Ploog Katja 2013. "Frontières. Question(s) de frontière(s) et frontière(s) en question: des isoglosses la mise en signification du monde", in Simonin, Jacky & Sylvie Wharton 2013 (eds.). Sociolinguistique du contact. Dictionnaire des termes et des concepts, Lyon, ENS Editions: 263–287.

Patriarca, Marco & Els Heinsalu 2009. Influence of geography on language competition, Physica A 388: 174.

Pike, Kenneth. 1948. Tone Languages. A Technique for Determining the Number and Types of Pitch Contrasts in a Language, with Studies in Tonemic Substitution and Fusion. Ann Arbor: University of Michigan Press.

Ross J. & A.P. Arkin, 2009. Complex systems: From chemistry to systems biology, PNAS 106, 6433 6434.

San Miguel, M., Eguiluz, V. M., Toral, R. and Klemm, K., 2005. Binary and multivariate stochastic models of consensus formation, Comput. Sci. Eng. 7: 6773.

Solé, R., Corominas-Murtra, B. and Fortuny, J., 2010. Diversity, competition, extinction: The ecophysics of language change, Interface 7: 16471664.

Stauffer, D. and Schulze, C., 2005. Microscopic and macroscopic simulation of competition between languages, Phys. Life Rev. 2: 89.

SSDSH 2011-16. Microrregión 13: Zona Mazateca, Secretaria de Desarrollo Social y Humano (SSDSH).

Steels, L. 2011. Modeling the cultural evolution of language, Phys. Life Rev. 8, 339356.

Schwartz, Diana 2016. Transforming the Tropics: Development, Displacement, and Anthropology in the Papaloapan, Mexico, 1940s1960s, PhD. Dissertation, University of Chicago.

Wichmann, S., 2008.The emerging field of language dynamics, Language and Linguistics Compass 2/3: 442.

Part III
Epilogue

Chapter 18
Epilogue

Dhruv Raina and Anirban Chakraborti

Between the Econophys-Kolkata conference organized in 2005 and Econophys-2015, we reckon that there has indeed been a widening of the agenda of the network of researchers and the themes being researched in the areas of econophysics and sociophysics. The participants at the last conference, the contributions to which appear in this volume, included economists, financial mathematicians, bankers and researchers located at different research institutions, computer scientists, mathematical physicists and mathematicians. And while economists and sociologists attended the meeting, their participation in the dialogue is still wanting. As just pointed out, thematically this conference resolved to widen the agenda of the network by moving from econophysics to econophysics and sociophysics. While earlier conferences too did engage with sociophysics, the research problematics were restricted to the sociophysics of markets and networks. The focus of research of econophysics over the past twenty years has thus been wealth distribution, stock markets and minority games, markets and networks, games and social choices, order driven markets, systematic risk and network dynamics, agent based models, and finally data driven models of market dynamics.

This conference extended some of the concerns of sociophysics to address complex social systems and phenomena that extended beyond market dynamics and networks, e.g., this involved examining the interaction and cooperative behavior among the extrovert and introvert agents and how the interaction evolves in time and determines the behaviour of the system [see the work of Dhar et al. in Sect. 13.1]. Sen presented in the conference her work based on the constrained Schelling model of social segregation [see Ref. Phys. Rev. E. **93**, 022310 (2016)]. Santhanam presented

D. Raina
Zakir Husain Centre for Educational Studies, School of Social Sciences,
Jawaharlal Nehru University, New Delhi 110067, India
e-mail: d_raina@yahoo.com

A. Chakraborti (✉)
School of Computational and Integrative Sciences, Jawaharlal Nehru University,
New Delhi 110067, India
e-mail: anirban@jnu.ac.in

F. Abergel et al. (eds.), *Econophysics and Sociophysics: Recent Progress and Future Directions*, New Economic Windows,
DOI 10.1007/978-3-319-47705-3_18

the mathematical modeling of seemingly complex phenomena like financial bubbles and crashes, based on the time series analyses of extreme events and record statistics in respect of empirical financial time series data [see Sect. 7.1].

Here, it was interesting to observe the frequent use of terminology from the social sciences such as "elitist" model or "egalitarian" model deployed within the formalism of network theory and not necessarily in ways that these terms are used as concepts in the social sciences. But this multiplicity is itself a reflection of the serious attempts to forge an interdisciplinarity driven by the compulsion of understanding complex social and socio-economic phenomena. In the 2010 Special volume on "Fifteen Years of Econophysics Research" [see Eds. B.K. Chakrabarti and A. Chakraborti, *Science and Culture* (Kolkata, India) **76 (9–10)** (2010)], there was an article written by Bertrand Roehner, where he had reviewed the evolution of the field in his address 'Fifteen years of Econophysics: Worries, Hopes and Prospects. There he highlighted the need to engage with social interactions and extend the methods of econophysics to demographic problems. He explained that the physicists' usual way of working has been to reduce complex phenomena into simpler phenomena. But he raised a question that while studying econophysics why should one make the effort of trying to break up complicated phenomena, when it is possible to handle them globally?

It appears that since then some headway has been made, and we will never know how much, unless bridges with the social sciences are forged. At stake are different ways of looking at theories, of the nature of models being developed, and how the models are to be interpreted. The sociologist Dipankar Gupta raised a number of interesting points about this divide in his inaugural address in the conference on 'Borders, Transgressions and Disciplinary Dynamics'. Interestingly enough the concept of borders and its 'twin concept' boundaries has in the recent past been the core theme of research in the social sciences posing problems for research on social and collective identity, demographic or census categories, immigration, cultural capital and membership, etc. But as Michèle Lamont and Virág Molnár point out in their piece on 'The Study of Boundaries in the Social Sciences', synthetic effects are still absent. In any case, it is evident that the engagement with boundaries is likely to illuminate a number of social processes that characterize apparently unrelated phenomena—and it is in this realm, perhaps that econophysics and sociophysics have much to offer in the near future.

Lightning Source UK Ltd.
Milton Keynes UK
UKOW06n1124060217
293708UK00002B/2/P